Determinants of Soil Loss Tolerance
ASA Special Publication Number 45

Proceedings of a symposium
sponsored by Division S-6
of the Soil Science Society of America
in Fort Collins, Colorado, 5–10 Aug. 1979.

Editorial Committee
B. L. Schmidt, Chmn.
R. R. Allmaras
J. V. Mannering
R. I. Papendick

Managing Editor
David M. Kral

Assistant Editor
Sherri Hawkins

Organizing Committee
J. V. Mannering, Chmn.
R. A. Gilkeson
G. O. Klock
W. C. Moldenhauer
R. D. Daniels
Leon Kimberlin

1982
Published by the
AMERICAN SOCIETY OF AGRONOMY
SOIL SCIENCE SOCIETY OF AMERICA
677 South Segoe Road
Madison, Wisconsin 53711

Copyright 1982 by the American Society of Agronomy and Soil
   Science Society of America, Inc.
   ALL RIGHTS RESERVED UNDER THE U.S. COPYRIGHT
   LAW OF 1978 (P.L. 94-553). Any and all uses beyond the
   limitations of the "fair use" provision of the law require written
   permission from the publisher(s) and/or the author(s); not
   applicable to contributions prepared by officers or employees of
   the U.S. Government as part of their official duties.

American Society of Agronomy
Soil Science Society of America
677 South Segoe Road, Madison, Wisconsin 53711 USA

Library of Congress Catalog Card Number: 81-70367.
Standard Book Number: 0-89118-071-0

Printed in the United States of America

Contents

Foreword v
Preface vii

1   Historical Perspective of Accelerated Erosion and Effect on World Civilization
     H. E. Dregne 1

2   Natural Erosion in the USA
     S. A. Schumm and M. D. Harvey 15

3   Rate of Soil Formation and Renewal in the USA
     G. F. Hall, R. B. Daniels, and J. E. Foss 23

4   Soil Erosion Effects on Soil Productivity of Cultivated Cropland
     G. W. Langdale and W. D. Shrader 41

5   Some Soil Erosion Effects on Forest Soil Productivity
     G. O. Klock 53

6   Determinants of Soil Loss Tolerance for Rangelands
     J. Ross Wight and F. H. Siddoway 67

7   Technology Masks the Effects of Soil Erosion on Wheat Yields —A Case Study in Whiteman County, Washington
     H. A. Krauss and R. R. Allmaras 75

8   Soil Loss Tolerance
     E. L. Skidmore 87

9   Current Criteria for Determining Soil Loss Tolerance
     D. E. McCormack, K. K. Young, and L. W. Kimberlin ... 95

10 Erosion Tolerance for Cropland: Application of the Soil Survey Data Base
     R. B. Grossman and C. R. Berdanier 113

11 Improved Criteria for Developing Soil Loss Tolerance Levels for Cropland
     Terry J. Logan 131

12 Economics of Soil Erosion Control with Application to T Values
     John F. Timmons and Orley M. Amos, Jr. 139

Foreword

The world population is expected to increase from approximately 4.5 billion in 1980 to 6.3 billion in the year 2000 if the present rate of population growth does not change. The world's food production must increase at the rate of 2% per year to maintain the same food supply/population ratio we have today, but even more would be needed to improve diets in developing countries. At the same time that increased food production is needed, millions of acres of land are being subjected to major soil erosion losses. This combination of events has led many to conclude that losses in soil productivity due to erosion are such that we will not be able to maintain current levels of agricultural production, much less meet the growing needs of the future.

Yet, in spite of obvious soil erosion losses, the productivity of much of our land has steadily increased during the past 40 to 50 years. This increased productivity in spite of erosion losses has supported the concept of tolerable rates of erosion for given soils below which there is no significant decrease in productivity.

While the concept of a tolerable rate of soil loss or T-value has been widely used, there are still many unanswered questions and concerns about the impact soil erosion may be having on our nonrenewable soil resources. More specifically, the question is raised as to whether soils can experience a certain rate of erosion without there being a corresponding decrease in productivity which has been offset by the application of new technology. It was because of these uncertainties that a symposium was held during the 1979 annual meetings of the American Society of Agronomy and the Soil Science Society of America. The objective of the symposium was to further examine the factors involved in the tolerable soil loss concept and either to suggest means of more accurately determining such tolerable loss levels or to identify areas of research needed to establish those levels for given soils. This publication reports the proceedings of the symposium. The contributing authors are highly qualified and their efforts along with those of the organizers and the editors are greatly appreciated.

C. O. Gardner, ASA President, 1982
R. G. Gast, SSSA President, 1982

Preface

Intolerable soil losses from soil erosion were undoubtedly recognized for centuries, but expressions of concern in the United States were infrequent until the late 1920 and early 1930 era. Bennett and Lowdermilk (1938) stated that soil loss from soil erosion was perhaps "the most potent single factor contributing to the deterioration of productive land." Out of the paradox between intolerable soil loss from man-accelerated erosion and the inevitable loss from geologic erosion came the idea of a tolerable soil loss. Early attempts to quantify soil erosion, also in about 1940, prompted the concept further, when in 1947 a soil loss tolerance in cropland was formally applied to some prominent soil types. At this time, the soil loss tolerance was defined as the "maximum average annual permissible soil loss without decreasing productivity" (Browning et al., 1947).

Widespread application of a soil loss tolerance (T-value) has occurred since 1962, when T-values were determined by the U.S. Soil Conservation Service for most of the major soil types in the United States. At this same time, the T-value was defined as "the maximum level of soil erosion that will permit a high level of crop productivity to be sustained economically and indefinitely" (Wischmeier and Smith, 1978). Throughout the development and use of the T-value, there has been disagreement about definition and intended use.

The T-value has been used since 1962 as a conservation planning guide. All soils in the USA have been assigned T-values ranging from 4.5 to 11.2 metric tons/ha per year (2 to 5 tons/acre per year). The magnitudes of the T-values are based on soil depth, prior erosion, and other factors affecting soil productivity. In the erosion control planning process, soil loss estimates for a particular site determined by the Universal Soil Loss Equation are compared with a T-value for that site. Any cropping and management combination for which the predicted erosion rate is less than the T-value may be expected to provide satisfactory erosion control. This strategy has been followed by the Soil Conservation Service in the USA since the early 1960s.

However, numerous factors in the last few years have coalesced to warrant a critical evaluation of the true meaning and use of the T value. These are: (1) increasing evidence of soil deterioration from soil erosion, (2) a recognition that technological advances may no longer be sufficient to counter productivity losses in severely eroded soil, (3) a transition from continued and unlimited expansion of cropland to one of cropland loss from urban expansion, (4) governmental policies of agricultural export to counter a negative balance of trade, and (5) increasing public concern over costs and public responsibility for effective conservation policies. Recent estimates indicate that over 27% of the cropland in the United States has erosion rates exceeding tolerable soil loss, that 23% of the cropland exceeds the maximum tolerable soil loss of 11.2 metric tons/ha per year (5 tons/acre per year), and that 10% exceeds 22.4 metric tons/ha per year (10 tons/acre per year) (CNI, 1978).

This background of ambiguous T-value definition, differing opinion of intended use, and more pressures on our limited soil resource prompted the Soil Science Society of America to hold a symposium, Determinants of Soil Loss Tolerance, at their 1979 annual meeting in Fort Collins, Colorado. This special publication is a collection of 12 papers given at the symposium, plus two papers solicited later to enhance coverage of the symposium. Objectives of this symposium and this publication are to:

a) review the perspective of accelerated erosion and its effect on world civilization;
b) review current guidelines and rationale for the determination and use of soil loss tolerance;
c) suggest improved criteria for determining soil loss tolerance values for cropland, forest, and rangeland; and
d) define areas of research needs to support and improve criteria for soil loss tolerance.

Although the original T-value definition was concerned with a physical limitation of the soil to produce forage and harvestable crops, this proceedings displays repeated, and generally unsuccessful, attempts to incorporate other damages into the T-value. Some of these are on-site damages related to nutrient loss or gully formation; while off-site damages proposed to be included are water quality and sedimentation, or social and economic considerations related to governmental policy. The symposium shows that these on and off-site damages are not readily incorporated into a single T-value. Although the symposium focused primarily on tolerable soil loss in a cropland system, range and forest lands also have a need for the control of soil loss. Existing T-value concepts and guidelines originated for cropland are not necessarily suitable for range and forest lands.

The participants at the Fort Collins Symposium possibly raised more questions than they answered. It is quite obvious from the review that there is no unanimity of opinion concerning even the philosophy of T-values. There was the agreement that the present system has major weaknesses based on lack of sufficient scientific data to first, adequately predict rates of soil formation, and second, to predict the effects of erosion on soil productivity. Certainly, research efforts need to be expanded in these two critical areas. There was, also, the agreement that to be workable, T-values must be acceptable socially and politically as well as scientifically.

LITERATURE CITED

Bennett, H. H., and W. C. Lowdermilk. 1938. General aspects of the soil-erosion problem. p. 581–608. *In* Soils and Men. 1938 Yearbook of Agriculture. USDA, Washington, DC.

Browning, G. M., C. L. Parish, and J. Glass. 1947. A method for determining the use and limitations of rotations and conservation practices in the control of erosion in Iowa. Agron. J. 39:65–73.

Conservation Needs Inventory. 1978. 1977 National Resource Inventories. Soil Conservation Service, United States Department of Agriculture, Washington, DC.

Wischmeier, W. H., and D. D. Smith. 1978. Predicting rainfall erosion losses—a guide to conservation planning. Agric. Handb. 537. USDA, Washington, DC.

Editorial Committee
B. L. Schmidt, Chm.
R. R. Allmaras
J. V. Mannering
R. I. Papendick

Chapter 1

Historical Perspective of Accelerated Erosion and Effect on World Civilization

H. E. DREGNE[1]

ABSTRACT

Accelerated erosion has made an impact, at one time or another, virtually everywhere in the temperate and tropical zones of the world that have been cultivated or heavily grazed or logged-over. The process has been slowed down or even halted in most of Europe, Canada, the United States, Australia, South Africa, South Korea, the Soviet Union, China, and Chile, and a few other countries. Erosion continues unabated in most of Africa, the Middle East, the Andes Mountains, and the Mediterranean countries. It is worsening rapidly in the hilly and heavily populated regions of southern Asia.

INTRODUCTION

Soil erosion is a relentless process that is nearly impossible to stop, usually difficult to control, and easily accelerated by man. While it may not be true that erosion has caused the downfall of nations, it can certainly be said that accelerated erosion has had—and is still having—adverse effects on human welfare. Massive erosion has massive effects, as witness the impact on the deep South of the horrendous gully erosion occurring 50 to 150 years ago (Bennett, 1939) and the effect on the southern Great Plains of the wind erosion of the 1930s (Joel, 1937). Even more important than the dramatic appearance of massive gullies and of dust storms that

---

[1] Horn professor and director, International Center for Arid and Semi-Arid Land Studies, Texas Tech. University, Lubbock, Texas.

Copyright © 1982 ASA, SSSA, 677 South Segoe Road, Madison, WI 53711. *Determinants of Soil Loss Tolerance.*

turn day into night is the insidious impact of seemingly small amounts of sheet erosion that may go on unnoticed for decades (Lyles, 1975; Willis and Evans, 1977). Lowdermilk (1948) described the effect of accelerated water erosion on the land and people of early civilization in the Old World. This paper will discuss man-made soil erosion and its consequences in the more recent past and at present.

Erosion Periods

Three periods of accelerated soil erosion have had a particularly acute impact on land resources. The first occurred 1,000 to 3,000 years ago, the second between 50 and 150 years ago, and the third in the last 30 years. Virtually every country in the world has gone through or is now experiencing a period of severe man-made erosion. Most of the developed countries have reduced or controlled accelerated erosion but the developing countries are only beginning to do so.

The first period of accelerated erosion, between 1,000 and 3,000 years ago, centered on the Mediterranean, the Middle East, and China. It was caused by excessive cutting of timber, the expansion of cultivation as population centers developed, and invasions by armies. That was the time when the cedars of Lebanon were destroyed, the terraces that the Romans had constructed in North Africa were abandoned, and the loessial hills of China were subjected to cultivation and massive erosion. After that first major period of land degradation, there was a hiatus, except in China, when wars and pestilence brought about an easing of pressures on land resources.

During the 19th and early 20th centuries, a second period of accelerated erosion occurred in the countries where there had been waves of European migration and the introduction of export-oriented agricultural economies. In the colonies where there was a considerable native population, Europeans frequently became the holders of extensive estates, forcing the indigenous people on to the least desirable land (Dresch, 1977). Increased soil erosion was a natural consequence of the emphasis on a marketing rather than a subsistence economy and the concentration of native populations on poor land.

The latest round of accelerated erosion started some time in the forepart of the 20th century and is continuing today, generally getting worse every year. It is a phenomenon associated with developing countries in which population numbers and land pressures are increasing rapidly. The worst affected regions are in Latin America, Africa, and southern Asia (Rapp, 1975; Grove, 1974; Butzer, 1974; Peralta, 1978). Nepal and Lesotho are two depressing examples of the results of the third major erosion period. It is easy to predict that there will be more examples in the years ahead.

Accelerated erosion is the result of two factors: improper management of productive soils and exploitation of marginal lands. Even the best soils cannot long tolerate having furrows run up and down the slope or being unprotected from the pounding impact of raindrops or losing the organic matter that binds soil particles together in stable aggregates. Such improper management accounts for much of the gully and sheet erosion that led to yield losses and land abandonment decades and centuries ago, due largely to ignorance of the problem and its solution. Knowledge of the problem and its solution has enabled erosion to be controlled and land improved wherever the will to do so exists and financial resources are adequate.

Exploitation of marginal lands had occurred out of ignorance of the consequences, greed, or the pressure of a greatly increased population. The classic example of ignorance of the consequences is the agricultural development a millenium ago on the loessial plateau of China. Loess produces a highly productive soil which, unfortunately, is also exceptionally erodible. The latter was unrecognized until gully formation became so bad that entire land surfaces simply disappeared into the Yellow River. Erosion control measures have begun to be instituted but the problem now is truly staggering and, to make matters worse, population pressures limit the management alternatives. A similar situation exists on the steep slopes of the formerly timbered land around the Mediterranean Sea.

Greed and ignorance have been responsible for a clear-cultivate-erode-abandon exploitative sequence in the past, but heavy population concentrations have forced that on people in recent years, especially in the humid tropics. Population pressures have also been responsible for cultivation of land on the fringes of the semiarid regions where drought and wind erosion hazards are high. The Sahel countries on the south side of the Sahara, Kenya, Iran, and Pakistan are examples of that unhappy situation.

Conservation and People

Soil conservation is an old and yet a new practice. It is old in that terracing to control water distribution and, incidentally, to minimize erosion has been practiced in Asia and Latin America for many centuries. On the other hand, soil conservation as a government activity designed to preserve the natural resources of a country is of 20th century origin. Most countries had no soil conservation agency before the 1930s, and many are still without them. Yet, the last 40 years has seen erosion halted, or at least slowed down, in developed countries and some developing countries where there has been a national determination to stop land degradation. In some cases, notably in Europe, formerly eroded land has actually been improved in recent decades.

The past 100 years has seen accelerated erosion put its mark on much of the land. It sometimes seems as though the only non-eroded soils are found in inhospitable climates or in inaccessible places. Lessons learned in one place about the harmful consequences of land abuse seem not be remembered by the pioneers in another place. It is interesting to speculate whether this failure to apply proven techniques is due to a cavalier attitude toward new land, the absence of adequate economic incentives, a belief that erosion would not become a serious problem, or a lack of understanding of the threat and its control.

The period of the late 1800s and early 1900s (the second erosion period) was a time when people and governments in countries that had experienced earlier waves of European migration came to realize that erosion had assumed alarming proportions. In the United States, warnings came from persons such as Edmund Ruffin, a pioneer in erosion control, and the directors of the Patent Office in the 1850s and later years, culminating in Hugh Hammond Bennett's dedicated efforts in the 1920s and 1930s. Officials in the former British colonies were in the vanguard in warning of the land abuse that was occurring in Australia, South Africa, Pakistan, India, East Africa, and elsewhere in that far-flung empire on which the sun never set. Commissions and inquiry boards noted the seriousness of the problem and sounded the alarm about the dire consequences if nothing were done to remedy it. By and large, nothing was done then or for many years after the warnings were first sounded.

Effective erosion control depends upon the individual farmer being aware of the damage erosion can do, having an incentive to protect an essential resource, and being able to obtain technical advise on control measures. In the United States, the incentive came when it no longer was possible to abandon gullied land and move to the virgin lands in the west. Awareness came slowly, and only when a few dedicated individuals talked and wrote incessantly about the importance to the farmer and the nation of soil conservation. The techniques—at least in principle—have been known for centuries. Government programs have perfected them and provided the technical assistance needed to put them into practice, and industry has made the tools. Much has been accomplished in a voluntary partnership of farmers, ranchers, and government, yet much remains to be done. In virtually all developing countries, there is awareness of the seriousness of accelerated erosion but frequently the incentives and the technical assistance are missing.

Impact of Past Erosion

The long-lasting effect of uncontrolled erosion on human welfare is amply demonstrated by the amount of effort that is being made today to control erosion that began a thousand or more years ago. China is expending tremendous energy in reducing the soil loss on the loessial highlands that contribute so much sediment to the Yellow River. The famous brigade that was held up, until recently, as the model for all Chinese agriculture ("In agriculture, learn from Dazhai") lies in the loessial highlands of Shanxi Province, where the challenges to agriculture are indeed great. The silt load in the Yellow River is so large that the hydroelectric plant below Xian has never been in full operation. Preventing disastrous floods in the east China plain watered by the Yellow River requires a never-ending dredging of the river channel and eternal vigilance in protecting the levees. The cost in human lives and energy of erosion on the watershed of just one river is incalculable.

Soil erosion in the Mediterranean countries continues to be a problem, with a few exceptions. The granary of the Roman Empire in North

Africa is making a slow comeback but it is disheartening to still see barren landscapes surrounding once-flourishing Roman cities. In the more arid parts of Algeria, Morocco, and Tunisia, the third historical period of erosion in the 20th century has brought about renewed land deterioration. Concern by national governments led to the development of an ambitious plan for a green belt of trees along the northern edge of the Sahara (United Nations, 1977). The Green Belt would extend—not continuously—from Morocco to Egypt. Its purpose is principally to stabilize sand dunes and sandy soils and to improve grazing and dryland farming in the 150 to 250 mm (5.9 to 9.8 inch) rainfall zone. Algeria has begun planting trees in a strip of land 20 to 40 km (12.4 to 24.9 miles) wide and 1,500 km (932 miles) long, a very costly effort to protect a dwindling soil resource.

Iran and most of the other Middle East nations have also entered into the third period of accelerated erosion before correcting the problems of earlier land degradation. The current erosion period is due to population increases—human and livestock—which have extended cultivation into marginal climatic areas that should have been reserved for grazing (Forestry and Range Organization, 1977) and have caused overgrazing of the remaining land (Pearse, 1970). Large sums of money are being spent to control gullies and stabilize sandy soils.

UNITED STATES

In what is now the United States, water and wind erosion were recognized as problems in the east even before the Revolutionary War (McDonald, 1941), yet later settlers who cultivated land in the south, midwest, and northwest were no better conservers of soil than the early easterners had been. Erosion was extended into the western rangelands in the later part of the 19th century when overgrazing by cattle and sheep followed the large expansion of livestock numbers to capitalize on the high prices of meat in eastern markets. Opening of the west by the railroads had a host of adverse environmental effects; increased erosion was just one of them.

It was not until the 1930s that anything significant was done to control soil loss. As long as there was new land to move to, most farmers allowed erosion to continue until the land was worn out. With some notable exceptions, accelerated erosion followed the advance of settlers across the North American continent, and few people worried about it. Even today, only modest progress has been made in applying conservation practices to the nation's land (General Accounting Office, 1977), and the greatest impetus for action comes from pressures to reduce non-point source pollution of water supplies rather than from the desire to conserve an irreplaceable resource. Table 1 shows that there was a major improvement in range condition on public domain in the United States from the 1930s to the 1950s but little change in the past 30 years. Table 2 indicates that cropland needing conservation treatment decreased from 1967 to 1977, whereas forest and rangeland requiring treatment increased.

Table 1. Trends in rangeland condition on U.S. federal land (Hadley, 1977).

Range forage trend	1930–1935	1955–1959	1975
		%	
Improving	1	24	19
Unchanged	6	57	65
Declining	93	19	16

Table 2. Non-federal land needing conservation treatment, U.S. (Anonymous, 1979).

Land use	1967	1977
	%	
Cropland	64	58
Forest land	62	67
Rangeland	71	75

Soil erosion control in the USA has been due principally to terracing, strip-cropping, residue management, and, most recently, minimum tillage. Crop rotations have not played a major role on commercial farms.

The rapid expansion in the use of center-pivot sprinkler irrigation systems to bring into cultivation land that was too sandy and too rough for surface irrigation has increased the potential wind erosion hazard, especially in the western Great Plains. If irrigation is ever discontinued, for reasons of economy or lack of water, there is sure to be a marked increase in wind erosion on those sandy lands due to the very slow re-establishment of a perennial grass cover on abandoned farm land (Costello, 1944). Should that happen during a drought period, the Great Plains will have another—but bigger—Dust Bowl on its hands.

AUSTRALIA

The Murray Mallee in the state of South Australia is located in the low precipitation, high-rainfall-variability margins of the humid regions of southern Australia (Williams, 1978). Annual rainfall, concentrated in winter, amounts to 250 to 380 mm (9.8 to 15 inches), the native vegetation is a eucalyptus shrubland, and the soils are sandy on the wide parallel ridges and clayey between the ridges. Wind erosion is serious wherever the plant cover is removed from the dune-like ridges. Sheep raising was the sole agricultural occupation prior to the introduction of wheat cropping at the beginning of the 20th century.

Land use developments in the Murray Mallee provide a classic example of how government support for land settlement, availability of improved machinery, and boundless faith in the ability of man to conquer the land could combine to bring short-term profits, followed by steadily decreasing soil productivity leading to farm abandonment and, finally, an uneasy accommodation between man and the environment. As it turned out, a carefully planned and orderly settlement of 26,000 km^2 (10,038 mi^2) of Mallee land was thwarted by a misplaced faith in the value of fallowing and the dust mulch.

Cultivation of the Murray Mallee was first promoted by the state in 1906. As early successes were noted during a time when wheat prices were high and the weather favorable, cultivation was expanded into the climatically marginal areas to the north. Wind erosion had been a problem from the beginning; it became intolerable when the inevitable dry periods occurred. Yields decreased steadily as erosion continued. By 1936, wind erosion was so great that many farmers simply abandoned their land. In 1939, the South Australian government put the Marginal Lands Scheme into operation to rescue the Mallee from complete disaster. Among the measures taken were the reduction of cropping to 1 year in three or four, replacing fallow with grass and legume pastures, increasing size of holdings, and integrating crop and livestock production. The remedial measures had brought about much improvement by the end of the 1950s.

The singly most important management change was to replace the wheat-fallow system with a wheat-pasture rotation. Wind erosion still occurs, but at a much reduced level. However, the threat of massive soil degradation occurring again is always lurking in the background. The narrow margin between profit and loss that farmers now face makes it tempting to shorten the rotation when wheat prices are favorable, despite the risk of increased sand movement. Resisting that temptation will not be easy.

Soil erosion is also a legacy of overgrazing. Wind erosion in the eastern rangelands and water erosion in the west and northwest have necessitated large private and public expenditures to revegetate scalds, reduce flooding on degraded watersheds, and install erosion-control structures. The overly optimistic estimates of carrying capacity of ranges were responsible for a sheep and cattle population that was 50 to 100% larger 50 years ago than it is today. Every pastoral region in Australia (and the United States) has gone through a similar phase of animal number increase and decrease: first there is rapid expansion to far beyond the carrying capacity, then there is a sudden fall in numbers when a drought hits, and finally subsequent stocking is at a more reasonable rate that can be sustained indefinitely, after the land has already been damaged (Condon, 1978).

AFRICA

The major part of the Republic of South Africa has been farmed by Europeans for about a century and a half. Soil erosion was recognized officially as a problem in a government report of 1922, and recommendations were made for action to stop further soil loss (Scott, 1951). The report was prepared by a commission charged with determining the cause of an apparent increase in the frequency of droughts. However, the commission concluded that there had been no change in the climate, and what was perceived as an increase in droughts was actually an acceleration in soil erosion and consequent land degradation. Over-grazing, the absence of any range management system, and bad agronomic practices were the causative factors in land degradation. In the late 1940s, it was estimated that the agricultural productivity of the nation had dropped by 25% in the previous 25 years. The Soil Conservation Act of 1946 provided for far-reaching changes in conservation methods and supplied loans and grants to enable farmers to change over to the use of conservation methods. Native reserves and protectorates were not covered by the legislation, and erosion has continued to worsen there. A 1955 report claimed that terrifying soil erosion was evident throughout South Africa (Kokot, 1955), despite the efforts that had been directed toward controlling it. One of the former British protectorates in southern African, the Kingdom of Lesotho, had the dubious distinction of being one of the most severely eroded countries in the world, with about half of all the cultivated land affected by sheet and gully erosion (Maane, 1975).

Water erosion has become serious during the past 30 to 50 years in the subhumid regions of Kenya, Tanzania, Ethiopia, Nigeria, the Ivory Coast, and other countries lying along the south side of the Sahara (Grove, 1974). Wind erosion has assumed major proportions closer to the Sahara in the semiarid dry farming lands of Mali, Mauritania, Senegal, Niger, Nigeria, Chad and the Sudan (Cloudsley-Thompson, 1974; Delwaulle, 1973; Depierre and Gillet, 1971).

Erosion, both wind and water, has been accelerated in nearly all of Africa during the 20th century (Jacks and Whyte, 1939; Bennett, 1975). There are many reasons, among them being the marked increase in human and livestock population resulting from the relatively recent provision of veterinary and human health services in rural areas and the pacification of warring tribes by colonial powers. Along with the increase in human and animal numbers came increased felling of trees, overgrazing, shortening of the fallow period between crops, expansion of cultivation into climatically marginal areas, and, inevitably, accelerated erosion. The dramatic Sahelian drought called the world's attention to land degradation in that part of Africa, but productivity had been dropping for many years before the drought. Little has been done to reverse the trend.

ASIA

The woodcutting, overgrazing, and destructive cultivation in the Middle East that caused devastation thousands of years ago (Lowdermilk, 1948), has spread in the intervening years until there is little land left that has not suffered man-made degradation. Pearse (1970) contends that rangeland deterioration (and erosion) has accelerated since 1950, primarily due to doubling or tripling of livestock numbers, extensive plowing of rangelands, sedentarization of nomads, firewood cutting, expansion of well drilling into formerly inaccessible areas, and better transportation facilities. Destruction of vegetative cover on sandy soils in Iran has led to increased wind erosion and to strenuous efforts to stabilize the dunes that have formed (Niknam and Ahranjani, 1976). The area of abandoned cultivated land in Iran has doubled in recent years and the number of livestock on grazing lands was estimated in 1964 to be 12 times the carrying capacity (Pearse, 1971).

Severe water erosion in the hills and dryland farming (barani) areas of northern Pakistan has brought demands for soil conservation since at least 1877 (Anwar, 1955). Gully erosion became so bad that in 1944 about

200,000 ha (494,000 acres) were said to be permanently destroyed. Wind erosion has also been serious. Together, they are responsible for an estimated 12,000 to 30,000 ha (29,640 to 74,100 acres) of land going out of cultivation each year in the Punjab, alone (Punjab Barani Commission, 1976). Erosion in the Himalayan watershed of the Jhelum River, if allowed to continue, will doom the huge Mangla Dam hydroelectric power complex to a short life (McVean and Robertson, 1969).

Excessive population growth and the consequent demands for crop land and wood are responsible for denudation of mountain slopes and greatly increased erosion in northern India, Nepal, and Indonesia in recent years (Brown, 1978). Flooding and silting of river beds have accompanied the accelerated erosion in a fashion reminiscent of the rise in the level of the Yellow River that finally caused it to break through its dikes and wreak havoc across the alluvial plain of north China in 1852.

About 27% of the arable land of the Soviet Union is severely or very severely eroded (Pryde, 1972). A massive program of shelterbelt construction in the 1930s has reduced the severity of wind erosion in the Ukraine but opening of the virgin lands of western Siberia and northern Kazakhstan has led to accelerated wind erosion in that wheat producing area.

South Korea is a refreshing—and unusual—example of a nation reversing soil erosion by undertaking a strong program of reforestation (Eckholm, 1979). The same success has, apparently, been achieved in the mountainous regions of China, although the awe-inspiring erosion in the loessial highlands goes on.

LATIN AMERICA

In Latin America, soil erosion has worsened over the past 20 to 40 years as, again, land pressure arising from population increase has brought more land into intensive use (Eckholm, 1976). At the same time, a growing concern for soil conservation has slowed down the rate of land degradation but has not halted it, as yet. Breaking up large estates in the altiplano of Bolivia has had an adverse effect on land productivity as cultivation—without conservation practices—has been extended onto hill slopes (Preston, 1969). It is ironic that the Incas and others had successfully constructed marvelous terracing systems on unbelievably steep slopes before the Spanish Conquest (Donkin, 1979). yet their descendants cannot protect soils on much gentler slopes.

As yet another legacy from the past, Chile is trying to repair the erosion damage done when soils on the slopes of the central valley were exploited to produce wheat for sale in California during the gold rush of 1849. Terraces and sod-forming forage crops have helped repair the damage but gullies and eroded hill tops are reminders that it takes only a short time to harm soils but it takes a long time to renew them.

Whereas water erosion has made life miserable for farmers in Mexico, Central America, Colombia, Chile and elsewhere, wind erosion plagues the western pampas of Argentine. A survey completed in 1948 showed that about 30% of the semiarid pampas had been severely damaged by wind erosion during the first half of the 20th century (Prego et al., 1971). Sand dune stabilization has become necessary in order to protect cultivated land from encroaching dunes.

EUROPE

Man-made erosion in Europe is pretty much under control in the northern countries, thanks mainly to forestation programs, but remains a problem in the Mediterranean countries, particularly Greece, Italy, and Spain (Tomaselli, 1977). Greece and Italy have a long history of accelerated erosion that stripped the thin layer of soil from the rocky slopes that abounded in those two countries (Lowdermilk, 1948). Spain is fortunate in having more favorable land resources.

Erosion and Crop Yields

Despite accelerated erosion that certainly is impairing the basic productivity of arable soils throughout the world, crop yields per unit of land are generally increasing. The exception is where primitive methods of agriculture are used.

Grain yields increased, worldwide, by about 40% between 1960 and 1974 (Walters, 1975), and the trend for all major U.S. grains was upward from 1950 to 1973 (National Research Council, 1977). Corn (*Zea mays* L.) and sorghum [*Sorghum bicolor* (L.) Moench] yields rose rapidly in the United States whereas wheat (*Triticum aestivum* L.) yields increased slowly over that 24-year period.

Evidence that loss of topsoil reduces soil productivity is clear-cut (Lyles, 1975). Yet, crop yields are increasing even though accelerated erosion continues to be widespread. The seeming paradox is a testimony to the value of agricultural research and extension in providing technological advances (improved plant cultivars, better pest control, water conservation, etc.) that keep us ahead of erosion losses. Intuitively, we know that this favorable situation cannot continue indefinitely. Sooner or later, soil productivity will decrease to the point where the damage will be irreversible economically. That point has already been reached on rangelands of the Navajo Indian Reservation and near El Paso, Tex. and on cultivated lands in the semiarid pampas of Argentina, the mountains of Nepal, and the hills of Lesotho, as well as in other places on a smaller scale.

How long it takes erosion to reduce land productivity to the economically irreversible state is an important question that cannot be answered unambiguously. A single storm that causes an undercut river bank to collapse can completely destroy land in minutes. Three consecutive years of drought and high winds were enough to cause formation of massive sand dunes and blowouts that ruined cultivated land in the Great Plains in the 1930s and 1950s. Sandy rangeland invaded by mesquite in southern New Mexico became nearly useless hummocky and blowout land in the 35 years between 1928 and 1963 (Buffington and Herbel, 1965). Ruination of grazing land generally is a slower process than ruination of cultivated land, primarily because vegetative cover is lost as soon as land is plowed.

EROSION EFFECTS ON CIVILIZATION

In most cases, economically irreversible erosion damage is confined to severely gullied land, to land where sheet erosion has exposed or nearly exposed bedrock, and to large hummocks and moving sand dunes. Given the proper circumstances, such severe erosion can occur in a few years but more commonly takes 30 or more years, judging from anecdotal evidence and the Australia experience (Williams, 1978).

The effect of sheet erosion on deep soils is more difficult to estimate. A uniform soil 2 m (6.56 ft) thick over bedrock and being eroded away at the rate of 23 metric tons/ha (20 tons/acre per year) (twice the allowable rate) would require over 1,000 years to be reduced to a soil 30 cm (12 inches) thick. A hilltop soil with a dense B horizon at 20 cm (8 inches) and suffering the same erosion rate would lose all its topsoil in just 136 years and probably would become uneconomic to farm in less than half that time.

Cost of Erosion

The cost of erosion, in terms of yield reductions, is difficult to determine. Based on data relating topsoil loss to yield reductions, just 2.5 cm (1 inch) of topsoil loss is enough to reduce U.S. wheat yields an average of 60,000,000 bushels/year (Dregne, 1978). That amount of soil loss is almost certainly less than the wheatlands have actually experienced.

There are no good data to use for calculating the cost of soil erosion in terms of lost crop yields, funds and energy devoted to repairing erosion damage, and human and animal health. Extrapolating estimates from one erosion plot to the entire earth's surface is a hazardous exercise in which to engage. About all that can be said is that the cost is very high even if the effects on crop yields are ignored. The latter cost, expressed as cropland value, was estimated by Stallings (1957) at $750 million/year for the United States. It would undoubtedly be much higher today.

Perhaps the most costly result of soil erosion is related to damage done by the soil particles that are dislodged and moved downwind or downstream. Wind erosion of sandy soils at planting time commonly forces the farmer to replant one or more times because seedlings have been sand-blasted or covered up. Replanting as many as three times is not particularly unusual in the southern High Plains USA and farmers in the Sahel region of Africa have been known to replant up to 12 times. Such replanting not only is costly in terms of time and money, it also had the effect of shortening the growing season, which can be disastrous.

Deposition of water-borne sediment contributes to flooding of lowlands, silting of reservoirs and irrigation canals, sedimentation of harbors, and elimination of nesting and spawning areas for fish. The cost of waterway dredging, alone, runs into the hundreds of millions of dollars annually. A considerable fraction—perhaps a third— of the estimated 3.6 billion metric tons (4 billion tons) of soil washed from U.S. soils each year reached major waterways, the remainder being deposited within a short distance of where it originates. East of the Mississippi, most of that silt probably represents accelerated erosion. In the west, the great majority is the result of geologic erosion, principally of shale soils.

CONCLUSIONS

Accelerated erosion has made an impact, at one time or another, virtually everywhere in the temperate and tropical zones of the world that have been cultivated or heavily grazed or logged-over. The process has been slowed down or even halted in most of Europe, Canada, the United States, Australia, South Africa, South Korea, the Soviet Union, China, and Chile, and a few other countries. Erosion continues unabated in most of Africa, the Middle East, the Andes Mountains, and the Mediterranean countries. It is worsening rapidly in the hilly and heavily populated regions of southern Asia.

The largest regions of arable land that have not suffered serious erosion are the tsetse fly zones in Africa south of the Sahara, the Amazon and Congo basins, the humid pampas of Argentina, and the southeast Asia mainland. The technology to reduce soil erosion to tolerable levels is available. What is needed is the national will to act.

The national will to act undoubtedly will determine whether soil erosion will be allowed to continue until it assumes the proportions of a catastrophe. In this regard, it is sobering to be reminded of the observation Huxley (1937) made about erosion in the British African colonies. He lamented that nothing was being done to control an obvious and widespread problem. He noted that erosion is, above all others, a problem designed for shelving: "It is thorny, it is packed with political dynamite, and it will always keep for another couple of years."

History shows that a crisis or near-crisis event usually is necessary to galvanize people and governments into concerted action. In the developed countries, those events have already occurred and soil conservation has become national policy and been given substantial financial support. In the developing countries where erosion is serious, governments generally are aware of it, but taking action requires technical and financial resources they do not have or have obtained only recently. We are a long way from resolving the problem but progress is being made, however slowly.

LITERATURE CITED

Anonymous. 1979. National Resource Inventories. Press release, Soil Conservation Service, U.S. Department of Agriculture, Washington, D.C.

Anwar, Abdul Aziz. 1955. Soil erosion in the Punjab. Publication No. 111, Board of Economic Inquiry. Lahore, Pakistan. 89 p.

Bennett, Charles F., Jr. 1975. Man and earth's ecosystems. John Wiley and Sons, Inc., New York. 331 p.

Bennett, H. H. 1939. Soil Conservation. McGraw-Hill Book Company, Inc., New York. 993 p.

Brown, Lester R. 1978. The worldwide loss of cropland. Worldwatch Paper 24. Worldwatch Institute, Washington, D.C. 48 p.

Buffington, Lee C., and Carlton H. Herbel. 1965. Vegetational changes on a semidesert grassland range. Ecological Monographs 35:139–164.

Butzer, Karl W. 1974. Accelerated soil erosion: A problem of man-land relationships. *In* Ian R. Manners and Marvin W. Mikesell (ed.) Perspectives on Environment, Association of American Geographers Publication No. 13. p. 57–78.

Cloudsley-Thompson, J. L. 1974. The expanding Sahara. Environmental Conservation 1:5–13.

Condon, R. W. 1978. Land tenure and desetification in Australia's arid lands. Search 9:261–264.

Costello, David F. 1944. Important species of the major forage types in Colorado and Wyoming. Ecological Monographs 14:107–134.

Delwaulle, J. C. 1973. Desetification de l'Afrique au sud du Sahara. Bois et Forets des Tropiques 149:3–20.

Depierre, D., and H. Gillet. 1971. Desertification de la zone Sahelienne au Tchad. Bois et Forets des Tropiques 139:3–25.

Donkin, R. A. 1979. Agricultural terracing in the Aboriginal new world. University of Arizona Press, Tucson, Arizona. 196 p.

Dregne, H. E. 1978. The effect of desertification on crop production in semi-arid regions. *In* Glen H. Cannel (ed.) Proceedings of an International Symposium on Rainfed Agriculture in Semi-Arid Regions, University of California, Riverside, California. p. 113–127.

Dresch, J. 1977. The evaluation and exploitation of the West African Sahel. Phil. Trans. Royal Society of London, B. 278:537–542.

Eckholm, Erik P. 1976. Losing ground. W. W. Norton & Company, Inc., New York. 223 p.

―――――. 1979. Planting for the future: Forestry for human needs. Worldwatch Paper 26. Worldwatch Institute, Washington, D.C. 64 p.

Forestry and Range Organization. 1977. Conservation Resources in Iran. Forestry and Range Organization, Ministry of Agriculture and Rural Development, Imperial Government of Iran. 26 p.

General Accounting Office. 1977. To Protect Tomorrow's Food Supply Soil Conservation Needs Priority Attention. General Accounting Office. Report to the Congress, CED-77-30, Washington, D.C. 59 p.

Grove, A. T. 1974. Desertification in the African environment. African Affairs 73:137–151.

Hadley, R. F. 1977. Evaluation of land-use and land-treatment practices in semi-arid western United States. Phil. Trans. Royal Society of London B. 278:543–544.

Husley, Elspeth. 1937. The menace of soil erosion. Journal of the Royal African Society 36:365–370.

Jacks, G. V., and R. O. Whyte. 1939. Vanishing lands. Doubleday, Doran, & Company, Inc. Reprinted by Arno Press, New York, 1972. 332 p.

Joel, Arthur H. 1937. Soil conservation reconnaissance survey of the southern Great Plains wind-erosion area. U.S. Department of Agriculture Technical Bulletin No. 556. Washington, D.C. 63 p.

Kokot, D. F. 1955. Desert encroachment in South Africa. African Soils 3:403–409.

Lowdermilk, W. C. 1948. Conquest of the land through seven thousand years. Soil Conservation Service, U.S. Department of Agriculture, S.C.S. MP-32. 33 p.

Lyles, Leon. 1975. Possible effects of wind erosion on soil productivity. Journal of Soil and Water Conservation 30:279–283.

Maane, Willem. 1975. Lesotho: A Development Challenge. The World Bank, Washington, D.C. 97 p.

McDonald, Angus. 1941. Early American Soil Conservationists. Soil Conservation Service, U.S. Department of Agriculture, Misc. Publication No. 449, Washington, D.C. 62 p.

McVean, D. N., and V. C. Robertson. 1969. An ecological survey of land use and soil erosion in the Wet Pakistan and Azad Kashmir catchment of the River Jhelum. Journal of Applied Biology 6(1):77–109.

National Research Council. 1977. World food and nutrition study. Commission on International Relations, National Academy of Sciences, Washington, D.C. 174 p.

Niknam, F., and B. Ahranjani. 1976. Dunes and development in Iran. Forestry and Range Organization, Ministry of Agriculture and Natural Resources, Tehran, Iran. 21 p.

Pearse, C. K. 1970. Range deterioration in the Middle East. p. 26–30. *In* M. J. T. Norman (ed.) Proc. XI International Grassland Congress, University of Queensland Press, St. Lucia, Australia.

––––. 1971. Grazing in the Middle East: past, present, and future. Journal of Range Management 24:13–16.

Peralta, P., Mario. 1978. Processos y areas de desertificacion en Chile continental. Ciencias Forestales 1:41–44.

Prego, Antonio J., Robert A. Ruggiero, Florentino Rial Alberti, and Federico J. Prohaska. 1971. Stabilization of sand dunes in the semi-arid Argentine pampas. p. 367–392. *In* William G. McGinnies, Bram J. Goldman, and Patricia Paylore (ed.) Food, fiber and the arid lands, University of Arizona Press, Tucson, Arizona.

Preston, David A. 1969. The revolutionary landscape of highland Bolivia. Geographical Journal, Vol. 135, pt. 1. p. 1–16.

Pryde, P. R. 1972. Conservation in the Soviet Union. Cambridge University Press, London. 301 p.

Punjab Barani Commission. 1976. Report of the Punjab Barani Commission. Government of the Punjab, Lahore, Pakistan. 386 p.

Rapp, Anders. 1974. A review of desertization in Africa—Water, vegetation, and man. Secretariat for International Ecology, SIES Report No. 1, Stockholm, Sweden. 77 p.

Scott, J. D. 1951. I. Conservation of vegetation in South Africa. p. 9–27. *In* Management and Conservation of Vegetation in Africa, Commonwealth Bureau of Pastures and field Crops Bulletin No. 41, Aberystwyth, Wales.

Stallings, J. H. 1957. Soil Conservation. Prentice-Hall, Inc., Englewood Cliffs, New Jersey. 575 p.

Tomaselli, Ruggero. 1977. The degradation of the Mediterranean maquis. Ambio 6:356–362.

Walter, Harry. 1975. Difficult issues underlying food problems. Science 188:524–530.

Williams, Michael. 1978. Desertification and technological adjustment in the Murray Mallee of South Australia. Search 9:265–268.

Willis, W. O., and C. E. Evans. 1977. Our soil is valuable. Journal of Soil and Water Conservation 32:258–259.

United Nations. 1977. Transnational Green Belt in North Africa. United Nations Conference on Desertification A/CONF. 74/25. 54 p.

Chapter 2

Natural Erosion in the USA

S. A. SCHUMM AND M. D. HARVEY[1]

ABSTRACT

The time required for soil development is intermediate between that required for nonrenewable mineral resources, metals, coal, gas, petroleum, and for renewable resources, forests, water, crops. For example, soils form naturally at rates of 0.5 to 0.02 mm a year. Natural erosion is highly variable in space and time ranging from essentially zero to an average maximum of 1.0 mm per year. Average man-induced erosion is 2.0 mm per year, which is far greater than natural rates of soil formation and natural erosion. Therefore, agricultural soils are being depleted, and they must be considered to be a nonrenewable resource.

INTRODUCTION

One way to evaluate the seriousness of the soil erosion problem in addition to the obvious effects, such as filled reservoirs and aggraded channels, is to attempt to evaluate the natural rates of soil formation and erosion and to contrast these with erosion rates induced by the activities of man. Unfortunately, in order to make this comparison, data from different disciplines with different perspectives must be compared. The agriculturalist is concerned with relatively short-term erosion from gently sloping cultivated fields, whereas the geomorphologist and hydrologist are usually concerned with erosion over long time spans, on much steeper slopes, or from entire drainage basins. The data presented herein generally relate to the latter situation.

---

[1] Professor and Research Associate, Dep. of Earth Resources, Colorado State University, Fort Collins, Colorado.

Copyright © 1982 ASA, SSSA, 677 South Segoe Road, Madison, WI 53711. *Determinants of Soil Loss Tolerance.*

Table 1.

Terms Relating to Pre-Agricultural Erosion		Terms Relating to Post-Agricultural Erosion	
Normal		Accelerated	
	Natural		Man-induced
Geologic		Historic	

Terminology

A statement about terminology seems appropriate. In Table 1, some of the terms used to describe erosion are listed, and we find them misleading. For example, the commonly used terms, normal erosion and geologic erosion are meant to imply pre-agricultural conditions of low erosion rates, whereas accelerated erosion and historic erosion imply greatly increased rates of erosion caused by man. Neither conception is correct because of the great variability of natural erosion both spatially and temporally at present and throughout geologic time. On the other hand, in many parts of the world, historic erosion is not accelerated, and it cannot be distinguished from pre-historic erosion. In agricultural areas historic erosion is accelerated, but much historic or accelerated erosion elsewhere is natural. Therefore, we prefer to use the terms natural erosion and man-induced erosion for normal and geologic erosion and for accelerated and historic erosion, respectively. For example, rapid rates of natural erosion occur in the badlands of South Dakota and Arizona, but man-induced badlands elsewhere are not natural.

Clear evidence of past arroyo cutting, gullying, and high sediment yields in the presettlement USA proves that natural erosion rates can be high. Ruhe and Daniels (1965) have documented situations in Iowa where natural erosion was greater than man-induced erosion. The important distinction for our purposes here is to identify natural rates of erosion for comparison with man-induced rates.

DENUDATION RATES

Unlike the agricultural engineer and agronomist who are concerned with erosion in a limited area, the geomorphologist is frequently concerned with the denudation of continents (Selby, 1974). The suspended sediment loads of large rivers have been used to estimate the denudation rates (the rates of uniform lowering of the area) of drainage basins as large as that of the Mississippi River, which Judson and Ritter (1964) calculated to be 50 mm per 1,000 years. They estimated the denudation rate for the entire USA to be 60 mm per 1,000 years. Obviously these figures average a great range of data. For smaller drainage basins where total sediment yields are measured, denudation rates can be on the order of several centimeters per year, and an average maximum rate of denudation has been estimated to be 1 mm per year (Schumm, 1963), for areas of about 4,000 Km^2.

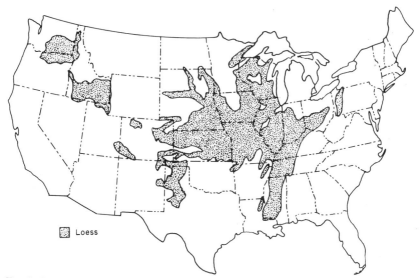

Fig. 1. Variation of natural denudation rates with precipitation. Dashed line suggests effects of extensive agricultural activity. Data were obtained from small watersheds with essentially natural vegetative cover (After Langbein and Schumm, 1958).

Natural erosion rates vary greatly with rock type and relief (Schumm, 1977, p. 21–25) within one climatic region, but they are also closely related to climatic controls. The relationship between natural rates of erosion and precipitation, for continental climates (Fig. 1) reveal that maximum natural erosion rates, for small drainage basins between 26 and 130 Km^2, occur in semi-arid regions (Langbein and Schumm, 1958). The form of the curve is largely dependent on vegetation. Under humid conditions vegetation protects the soil, but in arid regions there is insufficient rainfall and runoff to move large quantities of sediment. Therefore, at an intermediate rate of precipitation vegetational protection is low, yet there is sufficient runoff to move sediemnt out of the drainage basin. The broken line extension of the curve (Fig. 1) represents an upward shift of the right side of the curve, that would result from the removal of natural vegetation by agricultural activities.

The relation of Fig. 1 is helpful in developing an understanding of the effect of climate and man on erosion, but the records are short and frequently the quantity of sediment passing a gaging station does not necessarily reflect conditions in the sediment source area. Increasingly, evidence is becoming available that indicates that sediment does not move directly from fields and slopes to channels and thence out to the drainage basins. Rather, when upland erosion is great, sediment is stored in the valleys. For example, Costa (1975) calculated that only 34% of the sediment derived from agricultural erosion that commenced in the 1700s in Maryland has been carried out of the system. The detailed studies of Trimble (1975) and Mead and Trimble (1974) reveal the same patterns of sediment storage. A similar pattern has been documented in experimental studies of drainage-basin evolution (Schumm, 1977). High sediment production in a small experimental watershed was increased by lowering base level at the mouth of the basin. The main channel incised, and sediment yield increased dramatically, but it soon decreased as upstream sediments were stored in the valley. Through time, sediment yield varied significantly as sediment was alternately stored and eroded. These experiments provide further evidence that in areas of high sediment production, sediment yields will vary significantly through time.

Other studies reveal that once a drainage basin is in relative adjustment, there may be a considerable lag before sediment yields reflect upstream change. For example, a considerable reduction in upland erosion may not be detected down stream because the upland sediment is replaced by sediment eroded from the bed and banks of streams (Schumm, 1977, p. 325-326).

To summarize, the available information clearly demonstrates that natural rates of erosion are highly variable. They are very high in areas of high relief, of erodible rocks and soils, and in a semiarid climate. These rates can be further increased by man's activities, but the natural rates will also vary considerably through time. Catastrophic storm events, fires, and droughts can greatly accelerate natural erosion rates, and the fluvial system itself reacts complexly to natural and man-induced influences. Eroded sediment may be stored within the drainage network, and a lag in the downstream transport of this sediment is probable (Schumm, 1977).

SOIL FORMATION

Basic to the understanding of rates of soil formation are the rates at which rocks of different lithologies are weathered by chemical and physical processes under different climatic, topographic, and biologic conditions. Birkeland (1974) concludes that 10^5 to 10^6 years are required to develop weathered surfaces on granitic rocks, and larger periods are required for nongranitic clasts. Schumm and Chorley (1966) consider that weakly-cemented sandstones of the Colorado plateaus disintegrate in a matter of a few years, when freeze-thaw processes are active. Ruhe (1975) estimates that soil related to a Wisconsin surface weathered for 6,800 years, soil related to a late-Sangamon surface weathered less than 13,000 years, and soils related to a Yarmouth-Sangamon surface weathered for greater than 10^5 years. From this information, it is apparent that weathering and soil formation generally requires considerable time (Hall et al., 1981, Fig. 4).

The development of a mollic horizon (dark-colored, organic-rich A horizon) requires from 200 to 3,000 years, which is considered to be rapid as compared to the development of other soil characteristics (Birkeland, 1974). Trimble (1963) suggests that deeply weathered ultisols were formed in about 10^6 years in northwestern Oregon. On the other hand Franzmeier and Whiteside (1963) state that Spodosols take more than 3,000 years but less than 8,000 years to form. Kohnke and Bertrand (1959, p. 38) demonstrate that the total amount of soil formation, as expressed by the depth of the soil profile is a function of age. Their relationship indi-

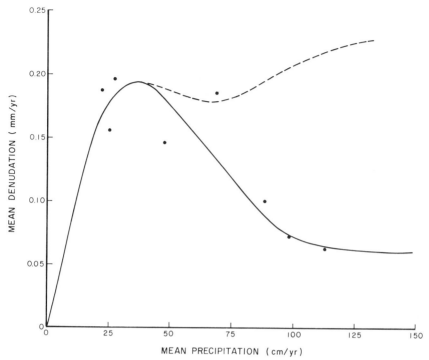

Fig. 2. Map of loess and other wind deposits in the USA (After Ruhe, 1975). The loess areas shown include bedrock outcrops, and alluvial deposits.

cates that 25 mm of soil may form in as little as 50 years (0.5 mm yr^{-1}), but that a soil depth of 2.5 m may require 150,000 years (0.02 mm yr^{-1}) to form. Pimental et al. (1976) consider that under ideal soil management conditions 25 mm of soil can form in 30 years (0.8 mm yr^{-1}). Under normal agricultural practice 25 mm of soil can form in 100 years (0.25 mm yr^{-1}). From this information it can be concluded that the natural erosion rate must have been less than 1 mm per year for the soils to have formed at all.

Fortunately, true soil development is not necessary for agricultural production, as is demonstrated by the highly productive loess deposits of the Midwest (Fig. 2). According to Frye et al. (1968) the Peoria loess of Wisconsin age provides the greatest thickness of loess, 20 to 30 m along the Platte, Illinois, middle Mississippi, and middle Missouri valleys. Soils have developed on these thick loess deposits, and they provide the basis for our highly productive midcontinent agriculture. These materials can be cultivated productively without true soil formation, as they are fertile deposits.

Let us contemplate Fig. 2, which is a map of the loess deposits of the USA and consider its implications. Large areas of our most fertile soils in Washington, Idaho, and the Midwest are developed from this loess, which was deposited during the glacial episodes of the Pleistocene. Accumulation of the material essentially ceased about 10,000 years ago, and although in many localities there are deep loess deposits, it is not forming in the Midwest at present. This loess, because of its method of emplacement, is readily eroded by wind and water and, in fact, is responsible for the highest known sediment yields in the world (Schumm, 1974).

Loess deposits, as well as alluvial deposits, can be agriculturally productive in spite of significant erosion, and they can continue to be productive until the loess or alluvium is removed by erosion from the underlying rock. When this occurs, productivity will cease if the underlying materials are not agriculturally productive. Both loess and alluvium are exotic materials that are derived from the erosion of earth materials. The loess was transported and deposited under conditions that generally no longer exist; therefore, it is a resource that is being permanently depleted.

SUMMARY AND PROGNOSIS

From a geomorphic perspective a wide range of erosion rates can be expected. When vegetational cover is significantly modified either naturally or by agricultural activity, erosion rates can be increased above the normal long-term average (1.0 mm yr^{-1}). Man-induced rates of erosion 10 times the natural rates have been reported following the modification of existing vegetation by agricultural practices, but Ruhe and Daniels (1965) also report a case of reduced erosion rates on loess following settlement in the Midwest (1.0 mm yr^{-1} to 0.5 mm yr^{-1}). The magnitude of the change depends on the nature of the material, the climate, and the morphology of the basins or slopes.

Although under certain ideal circumstances the rate of soil formation (2.5 mm yr^{-1}) is greater than the average rate of denudation (1 mm yr^{-1}), man-induced erosion (average 2 mm yr^{-1}) is significantly greater than natural rates of soil formation (0.5 to 0.02 mm yr^{-1}) and rates under normal agricultural practice (0.25 mm yr^{-1}).

If it is accepted that man-induced erosion occurs at a rate of 2 mm yr^{-1} then the thick loess deposits that are so valuable to the economy of the USA will be completely depleted in 10,000 to 15,000 years. Of course, the thinner deposits will be destroyed much sooner.

The geomorphologist has a longer term perspective than the pedologist or agronomist because he considers man's tenure to be brief and insignificant in terms of about 5 billion years of earth history. He also recognizes that man is selectively utilizing materials that have required long periods of time to form. Man's utilization of the soil resource by erosion can be placed in perspective by considering Table 2. Mineral resources, which were emplaced in excess of a billion years ago, are now being extracted and utilized. Energy resources, which were formed millions of years ago, are being severely depleted. Both are, therefore, considered to be nonrenewable resources that will not be replaced within man's existence. Forest resources are mainly renewable during the life span of man, and most crops and water are available on an annual basis, but some ground water is being mined, as is evidenced by falling water tables in the Southwest, where the resource is only renewable by transbasin diversions.

Table 2. Relative ages of natural resources.

	Resource	Age (years)	Replacement time (years)
Predominantly Non-renewable	Mineral resources gold-diamonds	10^9	billions of years
	Fossil fuels coal-petroleum	10^8-10^7	millions of years
	Soils	10^{3-4}	thousands of years
	Forest resources	10^3-10^1	tens to thousands of years
Predominantly Renewable	Water ground water surface water	10^3-10^0	1 to thousands of years
	(hydrologic cycle)	10^0	1 year
	Agricultural products crops	10^0	1 year

Clearly then, some resources are renewable within man's time scale and some are not, and soils fall within the category of nonrenewable resources. They are being mined and depleted just as mineral deposits and fossil fuels are being mined.

It seems unlikely, therefore, that a national policy of massive food exports can be maintained indefinitely, and in fact, such a policy, if pursued by the major food exporting nations will only hasten the rate of soil loss. Butzer (1974) succinctly puts the role of man-induced erosion in perspective as follows: "By the 1930s, when the efforts of the U.S. Soil Conservation Service began to take effect, the impact of man on the soil mantle, hydrology, and sedimentation had exceeded that of any natural climactically induced ruptures of equilibrium experienced in the southeastern USA during all of the Pleistocene times."

The general relations presented here are not a basis for the development of criteria for soil loss tolerance. Natural erosion rates are highly variable (Fig. 1) and, therefore, detailed investigations of natural erosion under many different conditions is required before comparison between natural and man-induced erosion can be made with confidence. However, we believe that the conception that most soils are a non-renewable resource is in itself a valuable contribution to consideration of this national problem.

ACKNOWLEDGMENT

Preparation of this paper was supported by the U.S. Army Research Office.

LITERATURE CITED

Birkeland, P. W. 1974. Pedology, weathering, and geomorphological research. Oxford University Press, New York. 285 p.

Butzer, K. W. 1974. Accelerated soil erosion: A problem of man-land relationships. p. 57-78. *In* J. R. Manners and M. W. Mikesell (ed.) Perspectives on environment. Association of American Geographers, Publ. No. 13.

Costa, J. E. 1975. Effects of agriculture on erosion and sedimentation in the Piedmont Province, Maryland, Geol. Soc. America, Bull., V. 86. p. 1281–1286.

Franzmeier, D. P., and E. P. Whiteside. 1963. A chronosequence of podzols in northern Michigan. I. Physical and chemical properties: Mich. State Univ. Agric. Exp. Stn. Quat., Bull., V. 46, p. 21–36.

Frye, J. C., H. B. Willman, H. D. Glass. 1968. Correlation of midwestern loess with the glacial succession. p. 3–21. In C. Bertrand Schultz, and J. C. Frye (ed.) Loess and related aeolian deposits of the world. Int. Assoc. for Quaternary Res., Lincoln, Univ. Nebraska Press.

Hall, G. F., R. B. Daniels, and J. E. Foss. 1981. Soil formation and renewal rates in the U.S. ASA and SSSA Spec. Pub. No. 45, Madison, Wis.

Judson, Sheldon, and D. F. Ritter. 1964. Rates of regional denudation in the United States. J. Geophys. Res., V. 69, p. 3395–3401.

Kohnke, Helmut and A. R. Bertrand. 1959. Soil conservation. McGraw-Hill, New York. 298 p.

Langbein, W. B., and S. A. Schumm. 1958. Yield of sediment in relation to mean annual precipitation. Am. Geophys. Union Trans, V. 39. p. 1076–1084.

Meade, Robert, and S. W. Trimble. 1974. Changes in sediment loads in rivers of the Atlantic drainage of the United States since 1900. Int. Assoc. Sci. Hydrol. Pub. 113. p. 99–104.

Pimental, D., E. C. Terhune, R. Dyson-Hudson, and S. Rocherau. 1976. Land degradation: Effects on food and energy resources. Science, V. 94. p. 149–155.

Ruhe, R. V. 1975. Geomorphology. Hougton Mifflin Co., Boston. 246 p.

―――, and R. B. Daniels. 1965. Landscape erosion-geologic and historic. J. Soil Water Conserv. V 20. p. 52–57.

Selby, M. J. 1974. Rates of denudation. N.Z. J. Geography, No. 56. p. 1–13.

Schumm, S. A. 1963. The disparity between present rates of denudation and orogeny. U.S. Geol. Survey Prof. Paper, 454-H. 13 p.

―――. 1974. Sediment yield of drainage systems. 15th Ed. Encyclopedia Brittanica. p. 473–480.

―――. 1977. The fluvial system. Wiley-Interscience, New York. 338 p.

―――, and R. J. Chorley. 1966. Talus weathering and scarp recession in the Colorado Plateau. Annals of Geomorphology, V. 10, No. 1. p. 11–36.

Trimble, D. E. 1963. Geology of Portland, Oregon, and adjacent areas. U.S. Geol. Surv. Bull. 1119. 119 p.

Trimble, S. W. 1975. Denudation studies: Can we assume stream steady state: Science. V. 188. p. 1207–1208.

Chapter 3

Rate of Soil Formation and Renewal in the USA

G. F. HALL, R. B. DANIELS, AND J. E. FOSS[1]

ABSTRACT

During the last 100 to 200 years man has greatly accelerated erosion on many agricultural fields. One result of this erosion has been the loss of an appreciable part of the rooting zone in soils having restrictive subsurface horizons. We believe that we should emphasize the maintenance of a favorable rooting zone. Obviously in soils shallow to rock or with other limiting layers that are not modified quickly, the erosion rate that is permissible should be lower than that in soils without restrictive horizons or layers. If we place emphasis on how rapidly we can develop a favorable rooting zone, we may be able to keep pace with the natural events that will take place on our landscapes in the next 2,000 to 5,000 years.

INTRODUCTION

It is often said that it takes a certain period of time to form an inch of soil. One of the most common figures used is 1 inch in 1,000 years. Seldom is any reason given for this figure but the basis for it may be Chamberlin's (1909) statement, referred to by Smith and Stamey (1965), to the effect that the mean rate of soil formation is less than 1 foot in 10,000 years based on observations since the glacial period. It is usually assumed that

---

[1] Professor, Agronomy Dep., The Ohio State Univ. and Ohio Agric. Res. and Dev. Ctr., Columbus, Ohio; formerly director, Soil Survey Investigations Division, USDA-SCS, Washington, DC, presently, visiting professor, Soil Science Dep., North Carolina State Univ., Raleigh, NC; and formerly professor, Agronomy Dep., Univ. of Maryland, College Park, Maryland, presently, professor, Soils Dep., North Dakota State Univ., Fargo, North Dakota.

Copyright © 1982 ASA, SSSA, 677 South Segoe Road, Madison, WI 53711. *Determinants of Soil Loss Tolerance*.

erosion rates exceeding that figure will remove the surface faster than rock or unconsolidated parent material are being weathered into soil material.

Does soil form in this fashion? A very significant aspect of soil formation is that it is made up of many facets; gains, losses, transformations, and translocations of different components taking place at different rates in different horizons (Simonson, 1959). For example, organic matter may be accumulating at the surface of the soil (Fig. 1) and carbonates and soluble salts may be leached to a considerable depth before diagnostic subsurface horizons form. Once leaching has taken place throughout several 10s of centimeters of the original material, some processes including clay migration take place throughout a relatively uniform depth (Stobbe and Wright, 1959), (Fig. 2). Degree of soil formation should be thought of as the extent of departure in properties at various depths from the properties of the original materials.

There seems to be a conventional wisdom that says that all soil formation is desirable and that without the orderly transition of soils through the genetic processes, soils will not be suitable for crop production or that at least they will become better suited for production with time. That this conventional wisdom is appreciably false is attested to by the birth of civilization on alluvial soils and the general fertility of these soils throughout the world.

The purpose of this paper is not only to review rates of soil formation and renewal but also to point out that some soil forming processes actually lead to less desirable conditions for crop production. On the other hand, it is important to note that without some of the geologic and pedologic processes, many of the soils of the nation would not provide a suitable medium for plant growth.

Concepts of Soil Age

The age of a soil can be thought of either as the absolute amount of time since some pedologically catastrophic event occurred or as the relative stage of development compared with some real or arbitrary standard. In the first concept we recognize a time zero when the soil parent material is first exposed to a given set of physical, chemical, and biological processes. This time zero may be some sudden event such as deposition of a thick deposit of new material, such as volcanic ash or fluvial deposits, or may be the result of earth slides or man's stripping of the pre-existing soil. It may also take the form of some longer termed change, such as geologic erosion, climatic change, change in vegetation induced by man's activities, or small incremental deposition by fluvial, colluvial, or aeolian processes.

In the time zero concept an attempt is made to determine an absolute age. The list of methods of putting an absolute age on an event is always increasing. The most commonly used dating technique is ^{14}C. This technique, based on the rate of decay of the unstable isotope of carbon, was developed in 1948. Although some of the assumptions associated with this

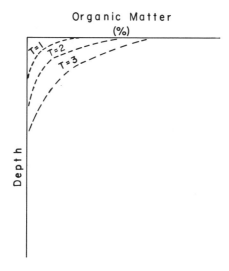

Fig. 1. Organic matter accumulation with time.

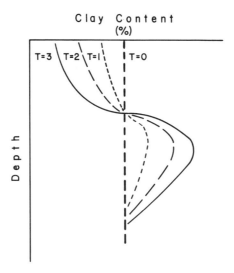

Fig. 2. Clay translocation and accumulation with time.

method have been questioned, it is still a good tool for determining the approximate time when a new cycle of soil formation began. It has a limit of 75,000 years before present (BP) or roughly from the start of the Wisconsin Pleistocene Period. Other radioisotopes have also been used for dating.

In addition to radioisotopes, a common method of establishing when soil development started is through the use of historic records. From these records, it is possible to determine when polders were drained, earthenworks were established, mountain glaciers retreated, and earth flows took place. Dating of soils by diagnostic artifacts has also been useful in pedologic-archeologic studies.

It is difficult to use absolute units of time in evaluating the age of a soil since soil formation may proceed at different rates because of differences or changes in one or more of the other factors. One approach is to use the stage or degree of development that has been reached in the soil. If a stage has been reached, provided that there were no other complications, it indicates a minimum time of soil formation for that stage. This approach assumes that we know the end stage and can identify all intermediate stages.

If we assume that there were no changes in factors, then:

Stage of soil formation = Intensity of the process × susceptibility of the material to soil formation × duration or absolute age.

Intensity of the process is a function of temperature, moisture, and organisms.

Susceptibility may be affected by
1) Permeability of the material,
2) Resistance of the material to weathering,
3) Carbonate content, and
4) Physical stability (topographic position).

The criteria to be used to determine the stage or degree of soil formation depend on which major diagnostic soil forming processes are operating, for example:

Mollisols—accumulation of organic matter in surface.

Alfisols—clay accumulation in subsoil.

Ultisols—clay accumulation in subsoil plus lower base saturation with depth.

Spodosols—accumulation of colloidal iron, Al, and humics in the subsoil.

Inceptisols—mineral weathering, color and/or structure in the subsoil.

Although these may be the dominant processes operating in the respective orders, other processes are active and could be used to evaluate development. Numerous other methods that have been used for evaluating degree or stage. These include:

1. Extent of clay formation or accumulation (Muckenhirn et al., 1949),

Relative age	Maximum content of colloid
1.0	29
1.7	44

2. Ratio of fine to coarse clay,
    The larger the ratio the older the soil, and
3. Depth of leaching,
    i.e., Illinoian — 3 m
        Wisconsinan — 1 m.

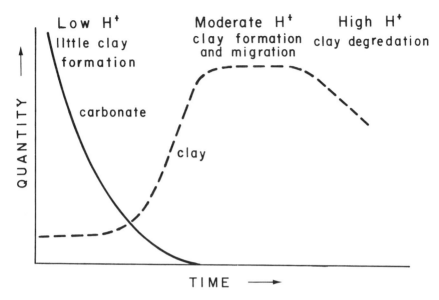

Fig. 3. Soil formation with time in a carbonate-rich medium.

Unfortunately there are a number of problems in using the degree of development as a criterion.

1. The environment changes. We know that over the past several thousand years there have been dramatic changes in climate in many areas with resultant changes in organisms. The changes have affected the kind and intensity of processes. For example in the central USA the vegetation has changed from coniferous forest to deciduous forest to grass in the past 10,000 years. In addition to this regional change in vegetation, there have been gradual local changes in the groundwater hydrology as a result of headward erosion into loess and glacial deposits in the region.

2. Often the sequence of processes or the intensity of processes changes with time as the result of changes in other processes. In some soils the sequence of clay formation, clay concentration, and clay degradation is related to the solution chemistry of the soil profile. This is illustrated in Fig. 3.

In addition to changes in clay content with time, the actual clay mineralogy will also change as a result of the changing soil chemistry. Cation exchange capacity (CEC), permeability, and other soil factors related to clay content also change.

3. On most land surfaces there are some additions of new material. These may be the result of aeolian (including aerosols), colluvial, or fluvial activities. The additions modify the existing physical and chemical processes of the soil.

4. Particulate material is lost through wind and water erosion.

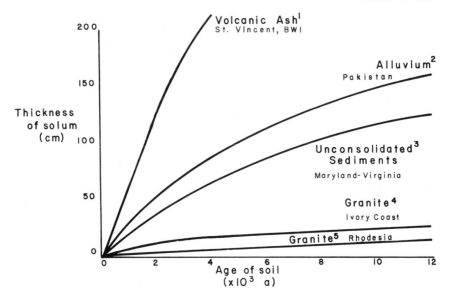

Fig. 4. Rate of soil formation in various geologic materials.

Soil Formation as Influenced by Geologic Materials

The mineralogy of the geologic material, along with the degree of consolidation, is a major influence on the rate of soil formation. Very siliceous geologic materials that have low porosity are very slow to weather and form a solum. Materials high in ferromagnesium minerals and with high porosity are on the other hand very susceptible to soil profile development. Figure 4 shows that consolidated geologic material, even under tropical climates, takes a long time to develop soil characteristics compared with unconsolidated materials. The figure also illustrates the climatic-soil material interaction that helps determine solum thickness.

Soil Age as Evaluated by Morphology

Soil profile development has long been utilized by pedologists to arrive at a stage of development or age, (Balster and Parsons, 1968; Gile, 1975; Gile and Hawley, 1968). Soil profile development in turn has been described by standard pedologic nomenclature (Soil Survey Staff, 1951).

Recently Bilzi and Ciolkosz (1977a, 1977b) developed a field morphology rating scale for evaluating this pedological development. The scale uses distinctness to compare horizons within a soil profile or to compare the C horizon with the solum for the evaluation of profile development. The morphological criteria used by Bilzi and Ciolkosz are: (1) Hue, value and chroma; (2) Textural class; (3) Grade, size, and type of struc-

Table 1. Rate of organic matter accumulation and formation of mollic epipedons.

Location	Criteria	Age (years)	Source
Alaska	A1 horizon	24–30	Ugolini, 1968
Iowa	Steady state O.C.	30	Ruhe et al., 1975
Wisconsin	A1 horizon	40	Nielsen and Hole, 1964
North Dakota	15 cm A horizon	50	Simonson, 1959
Alaska	Steady-state O.C.	60	Crocker, 1960
Iowa	2.6% O.C. in 0–31 cm	100	Hallberg et al., 1978
Oregon	Mollic epipedon	120	Parsons et al., 1970
Pennsylvania	4.1% O.C. in 0–8 cm	< 200	Bilzi and Ciolkosz, 1977b

ture; (4) Moist consistence; (5) Contrast of mottles; (6) Amount, thickness, and location of clay films, and (7) Soil horizon boundary conditions.

In four alluvial soils in Pennsylvania, these investigators found that the sequence of development parallels the known age (^{14}C) of the soils. They concluded that these development parameters could be applied in the humid temperature climatic region of northeastern United States but that additional morphological factors might be needed to evaluate soil profile development in other climatic regions.

Organic Matter Accumulation

Incorporation of organic matter in the surface of a parent material is usually considered to be the first indication of soil formation. The rate of incorporation is a function of climate and vegetation, type of parent material, and landscape position or topography. The time required for the formation of an A1 horizon is relatively short and has been studied by many researchers (Table 1).

Most of the studies have concentrated on soils formed under forest vegetation and a few on those formed under grass vegetation. In an early study, Crocker (1960) studying vegetation succession on glacial deposits concluded that organic matter accumulated rapidly for the first 60 years and then approached a steady state. In spoil material in North Dakota, Simonson (1959) observed that a 15-cm A horizon formed in 50 years. In a study of forest soils, Ugolini (1968) found that an A1 horizon developed in 24 to 30 years. Nielsen and Hole (1964) also studying soils developed under forest vegetation, showed that through a combination of earthworms and leaf litter an A1 horizon formed in 40 years. In studies of soil formation under grass vegetation in alluvial floodplains, several researchers (Ruhe et al., 1975; Parsons et al., 1970) found that Mollisols, or soils approaching Mollisols in organic matter accumulation, could form in 100 to 120 years.

A recent study of soil development under grass vegetation on loess parent material was conducted by Hallberg et al. (1978). They showed that in approximately 100 years the organic carbon content in the upper 10 cm of spoil material derived from loess was as high as that in an adjacent soil dated as being somewhat less than 14,000 years in age. The only

Table 2. Rate of clay migration and accumulation and rate of argillic horizon formation.

Location	Criteria	Age	Source
		years	
Iowa	Illuvial clay	100	Hallberg et al., 1978
Pennsylvania	Clay films	450	Cunningham et al., 1971
Iowa	Illuvial clay	1,100–1,800	Daniels and Jordon, 1966
Iowa	Clay films-argillic	<2,000	Dietz and Ruhe, 1965
Pennsylvania	Illuvial clay	2,000	Bilzi and Ciolkosz, 1977b
Iowa	Clay films	2,500	Parsons et al., 1962
Oregon	Clay firms-argillic	2,350–5,250	Parsons and Herriman, 1976
			Balster and Parsons, 1968
New Mexico	Argillic	>5,000	Gile and Grossman, 1968; Gile and Hawley, 1968

Table 3. Rate of structural development, cambic horizon formation and formation of other horizons.

Location	Criteria	Age	Source
		years	
Alaska	Color B	55	Ugolini, 1968
Pennsylvania	Cambic	200	Bilzi and Ciolkosz, 1977b
Pennsylvania	Cambic	450	Cunningham et al., 1971
Oregon	Cambic	550	Balster and Parsons, 1968
Iowa	Moderate structure	1,740–1,960	Parsons et al., 1962
Pennsylvania	Fragipan-like	2,000	Bilzi and Ciolkosz, 1977b
Alaska	Bir horizon	250	Ugolini, 1968

criterion that kept the soil from being classified as a Mollisol was an occasional light colored ped in the 10 to 30-cm depth.

All these studies suggest that organic matter can accumulate very rapidly under either forest or grass vegetation. Accumulations that can qualify as an A or A1 horizon take place in a matter of 10s of years and a steady state between gains and losses can be reached in a few hundred years.

Clay Accumulation

Translocation and accumulation of clay in the B horizons is generally acknowledged to take longer than organic matter accumulation, however studies tend to indicate that under ideal conditions the process may be relatively rapid (Table 2).

Parsons et al. (1962) in a study of soil genesis on an Indian mound in northeastern Iowa, determined that, in addition to a structural B horizon, clay films had formed in 2,500 years. In a study of soils developed on a landscape no older than 2,080 years, Dietz and Ruhe (1965) reported that a strong textural B horizon had formed in fine textured material deposited in swales. They found that on adjacent swells there were some well developed horizons but no clay films. Hallberg et al. (1978) studied soil development in leached loess spoil material under prairie vegetation receiving 86 cm of rainfall. They reported that there was some translocation and accumulation of illuvial clay and some fine silt in the subsurface horizons. Monona soils in western Iowa that are less than 1,800 years old have as much clay in the clay maximum as adjacent soils about 15,000 years old (Daniels and Jordon, 1966). The younger soils were on slopes ranging from 11 to 25%. Monona soils less than 1,100 years old have within 2 to 4% of the clay content of the older soils. None of these soils had clayskins. These studies suggest that under moderate rainfall a clay bulge can develop in a few thousand years.

Structural Development

Structural development is usually considered to be desirable for the enhancement of crop growth. If the initial material is massive or platy, the formation of soil structure normally improves the permeability of the soil. This improvement allows water to move into and through the soil more rapidly and additionally provides for planes of weakness along which plant roots can grow.

A cambic horizon and a color B horizon are usually thought to be the first stages of subsoil development. The rate at which a cambic horizon or color B horizon forms depends on all the soil forming factors. Some actual times required for these horizons to form are shown in Table 3. In an Alaskan soil Ugolini (1968) was able to identify a color B horizon after only 55 years of pedogenesis. The work from Oregon and Pennsylvania (Balster and Parsons, 1968; Bilzi and Ciolkosz, 1977b; Cunningham et al., 1971) all found cambic horizon development in soils less than 200 years of 550 years old. These are relatively short times pedologically. Parsons et al. (1962) reported a longer time but they also found indicators of clay migration which suggests that the soil they studied was beyond the stage of initial structural development.

Other Genetic Horizons

Numerous other types of genetic horizons may form as a result of the interaction of the soil forming factors (Table 3). Many of these have not been studied in enough detail to know the time required for formation. Many Spodosols are formed in late Wisconsin age glacial material, so it is apparent that less than 10,000 years is required to form a spodic horizon. Ugolini (1968) in his study of a soil in Alaska identified a Bir horizon in a soil approximately 250 years old. Soils with a 10 to 20-cm sequence of A1, A2, Bh horizons have been found on sandy dredge spoils that are less than 50 years old in North Carolina (J. Parnell, personal communication, UNC, Wilmington, 1972).

Fragipans are also formed in late Wisconsin age material. In 2000-year-old Pennsylvania alluvium, Bilzi and Ciolkosz (1977b) described a fragipan-like horizon. Some alluvial soils in Mississippi between 5,000 and 10,000 years oid have thick, strongly expressed fragipan horizons

(Earl Grissinger, personal communication, soil scientist, ARS, Oxford, Miss. 1979). Therefore it is presumed that formation of a fragipan requires more than 2,000 years but less than 5,000 years, if conditions are favorable. The degradation in these soils is greater than that in the Granada series in the adjacent upland.

Formation of Restrictive Horizons

As suggested earlier, initial soil genetic processes commonly lead to soil conditions that are more conducive to water movement and rooting than the original parent material. As the genetic processes continue, many of the soil horizons become less desirable for plant growth. The argillic horizon is desirable in very sandy soils because it helps to increase the available water capacity within the rooting zone. But in many medium to fine-texture soils, an argillic horizon can be a zone relatively impermeable to water and water may perch above it during periods of seasonally high rainfall.

Fragipan is another example of a horizon that is restrictive to movement and root growth. These pans commonly have a bulk density of 1.8 to 2.0 and may be 10 cm to more than 50 cm thick (Fritton and Olson, 1972; Yassoglou and Whiteside, 1960). Bulk density is lower in horizons both above and below these pan horizons.

Calcic and petrocalcic horizons formed in arid and semiarid regions tend to be both chemically and physically restrictive to root growth. In the extreme case of laminar precipitation of carbonate, the soil is effectively sealed to both water and roots (Gile et al., 1966).

In soil genetic processes the tops of these restrictive layers formed 10s of centimeters to a meter below the soil surface; with erosion or land leveling, these restrictive layers are often brought close to the surface. A number of studies have been conducted to determine the effect of these horizons on plant growth (Reuss and Campbell, 1961; Carlson et al., 1961; Batchelder and Jones, 1972). In Virginia, Batchelder and Jones (1972) reported that the subsoils are less fertile and have low available water storage capacity. Their studies showed, however, that with proper management the yields after 4 years were equal to the yields from the original surface soil. Reuss and Campbell (1961) had similar results in restoring productivity but cautioned that, although physical conditions were not a problem, soil structure must be restored. The restoration of desirable structure or the elimination of undesirable subsoil structure is often very difficult. In a study of New York pipeline trenches, Fritton and Olson (1972) found that the dense fragipan layer had been re-established in 11 years. They concluded that organic matter was essential for the retention of low bulk density.

In some soils the substrate below the soil horizons may be as dense and restrictive as the genetic soil horizons. Many of the unweathered glacial till deposits in central United States have a bulk density of 1.8 to 2.0 and are as restrictive to rooting as the fragipans. High density, restrictive horizons are commonly loosened by freezing and thawing cycles (Vorhees et al., 1978), but this may require a number of years. In the

Southeast the substrate commonly is Al saturated and exposure of these materials at the surface can result in poor growth of many crops (Gamble and Daniels, 1974).

Types and Rates of Additions

In many areas there are local additions of fluvial, colluvial, or aeolian materials by water, gravity, or wind. The amount of these additions depends on the landscape position and shape, the climate, and the vegetative cover. Fluvial contributions are most obvious along the major rivers of the country but are equally important for the soil renewal in the lower order tributaries. Much of the silt and sand-size fractions, lost from adjacent tracts, is deposited in alluvial positions and is not measured when sediment samples are taken from the major tributaries.

In most regions, the process of colluviation involves both gravity and water action. Movement of particles for short distances is common on all sloping landscape positions and the surface of the pedon represents a balance between surface gains and surface losses by lateral movement. It can be considered as a balance between the forces of erosion and the forces of renewal.

A less dramatic process of renewal is through aeolian additions. It attained dramatic proportions in the midcontinent during the 1930s and in the Sahel during the drought of the early 1970s, but most aeolian additions go on almost unnoticed. An early evaluation of renewal by aeolian additions was given by Free (1911). He concluded that airborne dust added at least 0.025 cm (about 3,000 kg/ha) annually to the area west of the Mississippi River. Smith and Stamey (1965) quote Chepil as stating that in southwestern Kansas the grass trapped at least 1.2 cm of aeolian material between 1946 and 1956. In a study of 12 sites encompassing an area from Kansas and Nebraska to Ohio, Smith et al. (1970) noted that the deposition of sand plus silt ranged from 75.7 kg/ha/month at North Platte to 16.8 kg/ha/month at Coshocton.

The size of dust storms has been studied in many parts of the world. They are often several thousand meters high (Prospero and Carlson, 1972; Van Hueklon, 1977; Lugn, 1968; Windom and Chamberlain, 1978) and the dust may be carried several thousand kilometers before it is deposited (Prospero and Carlson, 1972; Savoie and Prospero, 1976; Van Heuklon, 1977; Schutz and Jaehniche, 1976). By studying satellite images, it is often possible to determine the source of the dust and trace its movement (Prospero and Nees, 1977; Windom and Chamberlain, 1978).

Size of the aeolian material ranges from clay to fine silt. Handy (1976) stated that the largest particle that can be carried long distances is about 20 μm. This upper limit has been confirmed by a number of studies (Savoie and Prospero, 1976; Windom and Chamberlain, 1978; Schutz and Jaehniche, 1976; Van Heuklon, 1977).

Not all the aeolian additions are confined to the midcontinent near a Great Plains source area. Coen and Arnold (1972) attributed some of the clay minerals found in New York soils to an aeolian source. Tracking a February 1977 storm originating in the central part of the USA, Windom and Chamberlain (1978) traced it to the North Atlantic ocean off the coast of the Carolinas. They calculated that this one storm was capable of depositing 0.5 mm of clay-size material on the land or ocean surface.

The addition of increments of new material, whether from aeolian, fluvial, or colluvial sources, can cause many changes to occur in the soil. If the addition is very thick so as to essentially stop genetic processes in the buried soil, the new soil can be thought of as starting at time zero. More common than additions that completely bury the soil are the slow incremented additions that are characteristic of many landscapes. These slow additions, although often incrementally insignificant, cause significant changes in the soil through the addition of constituents. Carbonates or sulfates, for example, produce changes in soil chemical reactions and modify soil genetic processes. Man's contribution to these changes is becoming more apparent.

Soil Formation in Spoil Material

The vast acreages of mine spoil material have provided a unique opportunity for research on initial genetic processes in many kinds of parent material. The materials are present in many different climatic zones and on many different types of topography. Studying strip mine spoil has the advantage of knowing the starting time. It has the distinct disadvantage that often the material has a rather heterogeneous lithology. Another disadvantage is in the relatively short time over which most of these deposits have been subjected to the processes of soil formation.

The composition and particle size of spoil material depends on the geologic strata, the method of mining, and the length of time material has been exposed (Kohnke, 1950). Time zero for mine spoil is when the material is exposed. Weathering after exposure produces rapid changes in chemical and physical properties of mine spoil (Knudsen and Struthers, 1953) and glacial deposits (Crocker and Major, 1955). Although the spoil seems to be highly variable, it is not unlike the variability that can be found in a glacially deposited moraine. Kohnke (1950) found that within a few years shale, sandstones, and certain limestone fragments in mine spoil disintegrated into material less than 2 mm in size. Others have reported similar rates (Tyner and Smith, 1945; Knudsen and Struthers, 1953; Down, 1975). Down found that in the surface the greatest breakdown of fragments occurred in the first 21 years. Below a depth of 20 cm there has been almost no change in particle size in 178 years.

In the same study Down (1975) noted a significant decrease in pH for the first 21 years. From the 21st to the 55th year, there was no change in pH and then after the 55th year, pH increased significantly.

In a detailed study in Pennsylvania to study the effects of mining and reclamation on soils, Pedersen et al. (1978) compared mine spoil with adjacent natural soils. The spoils were youthful, and as expected, pedogenetic development was minimal. The mine spoils contained 50% more coarse fragments and had a bulk density that averaged almost 50% higher than that in the adjacent natural soils. Most of the roots in the mine spoil were located near the surface and were commonly found on coarse fragments or associated with the cracks and fissures near the coarse fragments. Roots were seldom seen in the massive C horizons. The absence of roots at depth was attributed to high density, lack of moisture, and lack of soil structure. Root development would be expected to increase with a decrease in density and an increase in porosity. The authors suggest that such changes can be expected as the result of freeze-thaw cycles and organic matter accumulation.

Strip mining, long considered an undesirable practice because of possible acid production, disruption of hydrologic systems, and undesirable esthetic consequences, can actually lead to a more desirable soil through improved slope and grade (Smith et al., 1974) and the destruction of compact soil layers (McCormack, 1974).

Comparison of Past and Present Erosion Rates

Erosion of our sloping farmland is a serious problem, but it is not unique to the present period of cultivation in either its widespread occurrence or rate. Most sloping landscapes in the world have been eroded at some time in the geologic past. Commonly this erosion is considered geologic, and usually is not thought to be excessively harmful. In fact, some benefits of geologic erosion can be expected if one considers the removal of horizons of low fertility or restrictive horizons and exposure of relatively fresh material as beneficial. An example of possible benefit is in Puerto Rico where oxisols of low fertility, but easily managed, cover old stable parts of the landscape, and mineral rich, thin Inceptisols occupy the recently eroded (geologically) valley sides (Daniels et al., in press).

Episodic erosion of hillslopes and deposition in valleys or low areas is common under a wide variety of climatic and landscape conditions (Butler, 1959; Antevs, 1952; Ruhe, 1964; Daniels et al., 1963; Bilzi and Ciolkosz, 1977b; Parsons et al., 1970; Walker and Ruhe, 1968; Gamble et al., 1970). These studies show that erosion has removed 3 to 4 feet or more of material from many valley sides (Ruhe and Daniels, 1965). Many of the soils on valley slopes and even on gently undulating till plains have been more or less severely eroded or received material in the last 5,000 years. In some places, the rate of erosion before disturbance by man was comparable with or exceeded that computed for the same areas under cultivation. Although there may be some validity to the idea that under geologic erosion the soil slowly sinks into the landscape at a rate nearly equal to that of erosion, in many areas erosion has been severe enough to remove the preexisting soils.

SUMMARY

Detailed soil-landscape studies in the past 30 to 50 years combined with reasonably accurate methods of dating events indicate that soil development can be much more rapid than originally thought. If these studies tell us anything, they point out that erosion and deposition, not uncommonly of catastrophic proportions, are part of the normal evolution of the earth's surface. These processes proceed with, without, or despite man's help. Soil profiles or horizons should not be considered sacred—they are destroyed and reformed too frequently in nature to allow us this luxury. The soil-landscape studies have led us to the idea that soils, for most crops requiring a well aerated environment, probably are most productive reasonably early in their development. Their maximum potential is reached after a certain level of organic matter and structural development has taken place but before the development of horizons that restrict root, air, and water penetration.

How can these ideas be applied to the problems of soil erosion on our cultivated soils? During the last 100 to 200 years man has greatly accelerated erosion on many agricultural fields. One result of this erosion has been the loss of an appreciable part of the rooting zone in soils having restrictive subsurface horizons. We believe that we should emphasize the maintenance of a favorable rooting zone. Obviously in soils shallow to rock or with other limiting layers that are not modified quickly, the erosion rate that is permissible should be lower than that in soils without restrictive horizons or layers. If we place emphasis on how rapidly we can develop a favorable rooting zone, we may be able to keep pace with the natural events that will take place on our landscapes in the next 2,000 to 5,000 years.

LITERATURE CITED

Ahmad, M., J. Ryan, and R. C. Paeth. 1977. Soil development as a function of time in the Punjab River Plains of Pakistan. Soil Sci. Soc. Am. J. 41:1162–1166.

Antevs, E. 1952. Arroyo cutting and filling. J. Geol. 60:375–385.

Balster, C. A., and R. B. Parsons. 1968. Geomorphology and soils, Willamette Valley, Oregon. Oregon State Agric. Exp. Stn. Spec. Rep. 265:8–22.

Batchelder, A. R., and J. N. Jones, Jr. 1972. Soil management factors and growth of *Zea mays* L. on topsoil and exposed subsoil. Agron. J. 64:648–652.

Bilzi, A. F., and E. J. Ciolkosz. 1977a. A field morphology rating scale for evaluating pedological development. Soil Sci. 124:45–48.

————, and ————. 1977b. Time as a factor in the genesis of four soils developed in recent alluvium in Pennsylvania. Soil Sci. Soc. Am. J. 41:122–127.

Butler, B. E. 1959. Periodic phenomena in landscapes as a basis for soil studies. CSIRO Soil Publ. No. 14.

Carlson, C. W., D. L. Grunes, J. Alessi, and G. A. Reichman. 1961. Corn growth on Gardena surface and subsoil as affected by application of fertilizer and manure. Soil Sci. Soc. Am. Proc. 25:44–47.

Chamberlin, T. C. 1909. Soil wastage. p. 75–83. *In* Proc. of a conference of governors in the White House, Washington, DC. 1908, U.S. Congr. 60th, 2nd Session, House Document 1425.

Coen, G. M., and R. W. Arnold. 1972. Clay mineral genesis of some New York Spodosols. Soil Sci. Soc. Am. Proc. 36:342–350.

Crocker, R. L. 1960. The plant factor in soil formation. Proc. 9th Pacific Sci. Congr. of Pacific Sci. Assn. 18:84–90.

————, and J. Major. 1955. Soil development in relation to vegetation and surface age at Glacier Bay, Alaska. J. Ecol. 43:427–448.

Cunningham, R. L., E. J. Ciolkosz, R. P. Matelski, G. W. Peterson, and R. W. Ranney. 1971. Characteristics, interpretations and uses of Pennsylvania soils: Armstrong County. Pennsylvania Agric. Exp. Stn. Prog. Rep. 316.

Daniels, R. B., F. H. Beinroth, L. H. Rivera, and R. B. Grossman. 1981. In press. Geomorphology and soils in an area of east-central Puerto Rico. SSIR No. 000.

————, and R. H. Jordon. 1966. Physiographic history and the soils and the soil, entrenched stream systems, and gullies, Harrison County, Iowa. USDA Tech. Bull. 1348.

————, M. Rubin, and G. H. Simonson. 1963. Alluvial chronology of the Thompson Creek Watershed, Harrison County, Iowa. Am. J. Sci. 261:473–487.

Dietz, W., and R. V. Ruhe. 1965. Soil development in less than 2,000 years in Iowa. Agron. Abstr. 57th Annu. Meet., Columbus, Ohio.

Down, C. G. 1975. Soil development on colliery waste tips in relation to age. J. Appl. Ecol. 12:617.

Foss, J. E. 1974. Soils of the Thunderbird site and their relationship to cultural occupation and chronology. In The Flint Run Paleo-Indian Occas. Publ. No. 1:66–83.

————, D. S. Fanning, F. P. Miller, and D. P. Wagner. 1978. Loess deposits of the Eastern Shore of Maryland. Soil Sci. Soc. Am. J. 42:329–334.

Free, E. E. 1911. The movement of soil material by the wind. U.S. Dep. Agric. Bur. of Soils Bull. 68.

Fritton, D. D., and G. W. Olson. 1972. Bulk density of a fragipan soil in natural and disturbed profiles. Soil Sci. Soc. Am. Proc. 36:686–689.

Gamble, E. E., and R. B. Daniels. 1974. Parent material of the Upper- and Middle-Coastal Plain soils in North Carolina. Soil Sci. Soc. Am. Proc. 38:633–637.

Gile, L. H. 1975. Cause of soil boundaries in an arid region: I. Age and parent material. Soil Sci. Soc. Am. J. 39:316–323.

————, and R. B. Grossman. 1968. Morphology of the argillic horizon in desert soils of southern New Mexico. Soil Sci. 106:6–15.

————, and J. W. Hawley. 1968. Age and comparative development of desert soils at the Gardner Spring radiocarbon site, New Mexico. Soil Soc. Am. Proc. 32:709–716.

————, F. F. Peterson, and R. B. Grossman. 1966. Morphological and genetic sequences of carbonate accumulation in desert soils. Soil Sci. 101:347–360.

Grube, W. E., Jr., R. M. Smith, J. C. Sencindiver, and A. A. Sobek. 1974. Overburden properties and young soils in mined land. In 2nd Res. Appl. Tech. Symp. on Mined-land Reclamation, Bituminous Coal Res., Inc., Monroeville, Pa.

Hallberg, G. R., N. C. Wollenhaupt, and G. A. Miller. 1978. A century of soil development in spoil derived from loess in Iowa. Soil Sci. Soc. Am. J. 42:339–343.

Handy, R. L. 1976. Loess distribution by variable winds. Geol. Soc. Am. Bull. 87:915–927.

Hay, R. L. 1960. Rate of clay formation and mineral alteration in a 4,000 year old volcanic ash soil on St. Vincent, BWI. Am. J. Sci. 258:354–368.

Knudsen, L. L., and P. H. Struthers. 1953. Stripmine reclamation research in Ohio. In R. J. Hutnik (ed.) Ecology and reclamation of devastated land. Gordon and Breach, N.Y.

Kohnke, H. 1950. The reclamation of coal mine spoils. Adv. Agron. 2:318–349.

Leneuf, N., and G. Aubert. 1960. Calculation of the rate of ferrolitization Int. Congr. Soil Sci. Trans. 7th (Madison, Wis.) 4:225–228.

Lugn, A. L. 1968. The origin of loesses and their relation to the Great Plains in North America. In B. C. Schultz, and J. C. Frye (ed.) Loess and related deposits of the world. Int. Assoc. Quaternary Res. Proc. 12:139–182.

McCormack, D. E. 1974. Soil reconstruction: For the best soil after mining. In 2nd Res. and Appl. Tech. Symp. on Mined Land Reclamation, Louisville, Ky. National Coal Assoc. Washington, DC.

Muckenhirn, R. J., E. P. Whiteside, E. H. Templin, R. F. Chandler, Jr., and L. T. Alexander. 1949. Soil classification and the genetic factors of soil formation. Soil Sci. 67:93–105.

Nielsen, G. A., and F. D. Hole. 1964. Earthworms and the development of coprogenous Al horizons in forest soils of Wisconsin. Soil Sci. Soc. Am. Proc. 28:426–430.

Owens, L. B., and J. P. Watson. 1979. Rates of weathering and soil formation on granite in Rhodesia. Soil Sci. Soc. Am. J. 43:160–166.

Parsons, R. B., C. A. Balster, and A. O. Neas. 1970. Soil development and geomorphic surfaces, Willamette Valley, Oregon. Soil Sci. Soc. Am. Proc. 34:485–491.

————, and R. C. Herriman. 1976. Geomorphic surfaces and soil development in the Upper Rogue River Valley, Oregon. Soil Sci. Soc. Am. J. 40:933–938.

————, W. H. Scholtes, and F. F. Riecken. 1962. Soils of Indian mounds in northeastern Iowa as bench marks for studies of soil genesis. Soil Sci. Soc. Am. Proc. 26:491–496.

Pedersen, T. A., A. S. Rogowski, and R. Pennock, Jr. 1978. Comparison of some properties of minesoils and contiguous natural soils. Interagency Energy/Environment R&D Program Report, EPA-600/7-78-162.

Prospero, J. M., and T. N. Carlson. 1972. Vertical and areal distribution of Saharan dust over the western equatorial north Atlantic Ocean. J. Geophys. Res. 77:5255–5265.

————, and R. T. Nees. 1977. Dust concentration in the atmosphere of the equatorial North Atlantic: Possible relationship to the Sahelian drought. Science 196:1196–1198.

Reuss, J. O., and R. E. Campbell. 1961. Restoring productivity to leveled land. Soil Sci. Soc. Am. Proc. 25:302–304.

Ruhe, R. V. 1964. Landscape morphology and alluvial deposits in southern New Mexico. Ann. Assoc. Am. Geogr. 54:147–159.

————, and R. B. Daniels. 1965. Landscape erosion-geologic and historic. J. Soil Water Conserv. 20:52–57.

————, T. E. Fenton, and L. L. Ledesma. 1975. Missouri River history, floodplain construction, and soil formation in southwestern Iowa. Iowa Agric. Home Econ. Exp. Stn. Res. Bull. 580:738–791.

Savoie, D., and J. M. Prospero. 1976. Saharan aerosol transport across the Atlantic Ocean: Characteristics of the input and output (abs.). Am. Meteorol. Soc. Bull. 57:145.

Schutz, L., and R. Jaehniche. 1976. Mineral dust in the NE trade wind region of the Atlantic Ocean (abs.). Am. Meteorol. Soc. Bull. 57:145.

Simonson, R. W. 1959. Outline of a generalized theory of soil genesis. Soil Sci. Soc. Am. Proc. 23:152–156.

Smith, R. M., W. E. Grube, Jr., T. Arkle, Jr., and A. Sobek. 1974. Mine spoil potentials for soil and water quality. Environ. Prot. Technol. Series. EPA 670/2-74-070. U.S. Environ. Prot. Agency, Cincinnati, Ohio. 303 p.

————, and W. L. Stamey. 1965. Determining the range of tolerable erosion. Soil Sci. 100:414–424.

————, P. C. Twiss, R. K. Krauss, and M. J. Brown. 1970. Dust deposition in relation to site, season, and climatic variables. Soil Sci. Soc. Am. Proc. 34:112–117.

Sobek, A. A., R. M. Smith, W. A. Schuller, and J. R. Freeman. 1976. Overburden properties that influence minesoils. *In* 4th Symp. on Surface Mining and Reclamation, National Coal Assoc. p. 153–159.

Soil Survey Staff. 1951. Soil survey manual. USDA Handb. No. 18.

Stobbe, P. C., and J. R. Wright. 1959. Modern concept of the genesis of Podzols. Soil Sci. Soc. Am. Proc. 23:161–164.

Tyner, E. H., and R. M. Smith. 1945. The reclamation of the stripmined coal land of West Virginia with forage species. Soil Sci. Soc. Am. Proc. 10:429–436.

Ugolini, F. C. 1968. Soil development and alder invasion in a recently deglaciated area of Glacier Bay, Alaska. *In* J. M. Trappe, J. F. Franklin, R. F. Tarrant, and G. M. Hansen (ed.) Biology of alder. Pacific Northwest Forest and Range Exp. Stn. Forest Service, USDA, Portland, Oregon.

Van Heuklon, T. K. 1977. Distant source of 1976 dustfall in Illinois and Pleistocene weather models. Geology 5:693–695.

Vorhees, W. B., C. G. Senst, and W. W. Nelson. 1978. Compaction and soil structure modification by wheel traffic in the northern corn belt. Soil Sci. Soc. Am. J. 42:344–349.

Walker, P. H., and R. V. Ruhe. 1968. Hillslope models and soil formation: II. Closed system. p. 551–560. *In* Trans. 9th Int. Congr. Soil Sci. Adelaide.

Windom, H. L., and C. F. Chamberlain. 1978. Dust-storm transport of sediments to the North Atlantic Ocean. J. Sediment. Petrol. 48:385–388.

Yassoglou, N. J., and E. P. Whiteside. 1960. Morphology and genesis of some soils containing fragipans in northern Michigan. Soil Sci. Soc. Am. Proc. 24:396–407.

Chapter 4

Soil Erosion Effects on Soil Productivity of Cultivated Cropland[1]

G. W. LANGDALE AND W. D. SHRADER[2]

ABSTRACT

Soil erosion always increases the cost of crop production and causes potential environmental hazards as well as human suffering. Erosion of soils by water reduces crop yields principally through the loss of nutrients and available water. Exposed subsoils caused by severe soil erosion also exhibit many adverse properties with respect to soil management for economic crop production.

Agronomic implications of soil erosion by water in the United States have been derived mainly from limited research on Mollisols, Alfisols, and Ultisols. Because cultivated Ultisols of the southeastern USA are thinner and suffer problems associated with subsoil acidity, crop yield reductions appear more permanent and difficult to restore. The permanency of soil erosion on crop yield reductions on many Mollisols soils appears ephemeral, because only additional quantities of N, occasionally P, and micronutrients are required to restore crop yields.

Additional research is urgently needed to quantify crop yield losses associated with soil erosion and reduce the cost of restoring crop production to an economic competitive level on eroded landscapes. Research of this nature would also provide insights for controlling unacceptable soil erosion levels.

INTRODUCTION

Soil erosion affects crop production principally by reducing nutrient supply, water infiltration, and soil water-holding capacity. Adverse soil tilth and aeration may accompany progressive soil erosion. Surface con-

---
[1] Contribution from Southern Piedmont Conservation Research Center, USDA-ARS, Watkinsville, GA 30677 and the Dep. of Agronomy; Journal Paper No. J-9600 of the Iowa Agric. and Home Econ. Exp. Stn., Ames, IA 50011.
[2] Soil scientist, Southern Piedmont Conservation Research Center, USDA-ARS, Watkinsville, GA 30677; and professor of agronomy, Iowa State University.

Copyright © 1982 ASA, SSSA, 677 South Segoe Road, Madison, WI 53711. *Determinants of Soil Loss Tolerance.*

figuration of the landscape is also objectionably altered with severe soil erosion. Agronomic implications of soil erosion research in the United States have been derived mainly from lands used for the production of corn, soybean (*Glycine max* L.), cotton (*Gossypium hirsutum* L.), and small grains, which are all annual crops.

Early studies on soil productivity and soil erosion concentrated on diminished nutrient supply, because almost all of the plant available N in the soil is in the form of soil organic matter, usually concentrated in the surface 4 to 12 inches (10 to 30 cm). Also, 50% of the plant-available P is usually in the organic form (Black, 1968).

Numerous long-term studies in the United States from about 1935 to 1950 amply document a trend of reducing crop yield (Adams, 1949; Copley et al., 1944; Hays et al., 1948; Latham, 1940; Musgrave and Norton, 1937; Smith et al., 1945; Whitney et al., 1950). Yields of row crops studied during this period declined drastically on all soils as the surface soil was lost by erosion unless soil nutrients, organic matter, and occasionally water were intensively supplied. Universally, these studies suggest that when the soil surface was lost, the supply of N and P was drastically reduced and crop yields declined. Fertility inputs and crop yields reported for these studies were much lower than current crop production even on eroded soils. Where topsoil depths were less than 30 cm, crop yields on severely eroded surfaces were reduced 20 to 50%.

Studies with only a goal to provide vegetative cover for drastically disturbed or eroded nonagricultural lands support findings related to agricultural lands (Bennett, 1939; Parks et al., 1967; Peperzak et al., 1959; Richardson and Diseker, 1961; Vanderlip, 1962[3]). Numerous experimental sites on roadside backslopes which exposed calcareous loess and glacial till subsoils suggested that nitrifiable N and available P exerted the largest positive influence on plant growth. A high percentage of either sand or clay in the backslopes in Iowa reduced vegetative cover; possibly because of adverse water relations and supply. When rates of complete fertilizers were used, plant growth was achievable to control soil erosion. On drastically disturbed kaolinitic soils of the southeastern USA, mulches, lime, fertilizers, and rhizosphere activity are often required to achieve sufficient vegetative cover to control soil erosion. Kudzu (*Pueraria thunbergiana*) and native pines (*Pinus echinata* L.) are usually grown more successfully on these lands (Bennett, 1939).

Some rangelands and Pacific Northwest croplands are susceptible to rapid irreversible soil erosion conditions. Attention will be given to these lands in other chapters of this publication.

Shrader et al. (1963) stated that the onsite effects of soil erosion vary with the remedial measure as well as the variation in soil properties. Winters and Simonson (1951) provide an excellent treatise of desirable and undesirable properties of exposed subsoil as they affect plant growth. Some knowledge of exposed subsoil properties is necessary even to suggest remedial treatment for vegetative cover purpose. Individual crops have been known to respond differently to eroded soils for many decades in the

---

[3] Vanderlip, R. L. 1962. Interrelationships of mulches and fertilizers in erosion control on highway backslopes. Unpublished M.S. Thesis, Library, Iowa State University, Ames, Iowa.

United States (Adams, 1949; Baver, 1950; Buntley and Bell, 1976). The objective in this paper is to evaluate the current yield reduction extent and permanency of soil erosion east of the Rocky Mountains.

WHY SOIL EROSION REDUCES CROP YIELD

Land forming, generalized functions relating plant response to measured soil factors, or some combination of these studies provide the best insights to describe how soil erosion affects crop yields. Soils described in these studies are not necessarily subject to severe erosion. Henao[4] evaluated the effects of 95 soil factors on corn yields with multiple regression techniques over a wide range of soil conditions in 17 Iowa counties. Plant-available water-holding capacity of the soil was highly correlated with corn yields. Soil erosion estimates were also needed to explain corn yield variation. Soil erosion had an overall negative effect on yields, but there was no determination of specific soil effects, such as drainage characteristics, depth of solum, or clay content. This study included such soils as Ida (fine-silty, mixed, mesic Typic Udorthents) and Monona and Marshall (fine-silty, mixed, mesic Typic Hapludolls) which form in the medium textured, deep loess deposits, where only a small decrease in yields due to erosion would be expected under modern conditions of high fertilizer use. Included also were such soils as Shelby (fine-loamy, mixed, mesic Typic Argiudolls) or Seymour (fine, montmorillonitic, mesic Aquic Argiudolls) where dense subsoils would markedly decrease yields.

Thomas and Cassel (1979) measured changes in physical and chemical factors of the root zone caused by land forming Atlantic Coastal Plain soils. Regression methods were used to account for corn yields produced by these soil factors. The soil variables ranked in order of declining importance were (1) A horizon thickness, (2) plant available P, (3) bulk density, (4) available water-holding capacity, (5) organic matter content, and (6) plant-available K. This study included Goldsboro (fine-loamy, siliceous, thermic Aquic Paleudults), Lynchburg (fine-loamy, siliceous, thermic Aeric Paleaquults), and Rains (fine-loamy, siliceous, thermic Typic Paleaquults) soils. Surface soil thickness was important in this study because the subsoil (sandy loam to sandy clay loam textures) was high enough in exchangeable Al to limit root growth of most plants (Table 1). Webb and Bear[5] show that soil thickness of Edina silt loam (fine, montmorillonitic, mesic Typic Argialbolls) is of less importance in southern Iowa.

In Texas soils, Eck et al. (1965), Heilman and Thomas (1961), and Thomas et al. (1974) showed that low mineralizable N and available P limited plant growth in land-leveling studies. Their study soil sites were

---

[4]Henao, J. 1976. Soil variables for regressing Iowa corn yields on soil management and climatic variables. Unpublished Ph.D. Dissertation, Library, Iowa State University, Ames, Iowa.

[5]Webb, J. R., and C. Bear. 1972. Unpublished data from Southern Iowa Experimental Farm, Iowa State University, Ames, Iowa.

Table 1. Effect of surface soil thickness on corn yields in southern Iowa and eastern North Carolina.

Surface soil thickness	Average corn yields	
	Southern Iowa†	Eastern North Carolina‡
cm	q/ha (grain)	q/ha (total dry matter)
26	61	59
31	65	--
37	65	--
39	--	74
43	67	--
45	--	91
49	67	--
55	65	--

† Webb and Bear, 1972. Edina silt loam; a study associated with land forming disturbances.
‡ Thomas and Cassel, 1979. Goldsboro loamy sand, Lynchburg fine sandy loam, and Rains loamy sand, respectively.

Pullman silty clay loam (fine, mixed, thermic Torrertic Paleustolls) and Hidalgo sandy clay loam (fine-loamy, mixed hyperthermic Typic Calciustolls) with clay and clay loam subsoil textures, respectively. These soils are not necessarily subject to severe soil erosion. However, relationships associated with mechanically truncated soil properties to plant growth are important to elucidate the effects of soil erosion on soil productivity.

Deleterious effects of plant nutrient imbalances, soil texture, and bulk densities are not the only adverse conditions produced by exposing subsoil. Black and Greb (1968) indicated that exposing a Weld silt loam (fine, montmorillonitic, mesic Aridic Paleustolls) subsoil changes the reflectance (color) which causes a temperature interaction that may indirectly affect soil water storage and ultimately crop yields. Exposed fine textured subsoils also decrease water infiltration (Batchelder and Jones, 1972). Almost all soil researchers and farmers elude to tilth, surface configuration, and poor plant population problems on eroded soils. Organic matter, cations, and kind and amount of clay are the principal factors affecting soil aggregation or tilth. Olson (1977) discussed the soil tilth problem as it affected machinery performance and sparse plant stands on exposed glacial till.

CROP YIELD REDUCTION ON ERODED SOILS

AVAILABILITY OF SOIL EROSION-SOIL PRODUCTIVITY DATA

Reliable crop yield data on eroded soils with a noneroded comparison treatment is difficult to find in the literature. Randomized statistical field-plot designs are not useful tools for measuring crop yield variations on eroded landscape because the random-spatial nature of water erosion. Because of increasing crop yields caused by new technologies during the past 30 years in the United States, intensive soil erosion-soil productivity research accomplished between 1935 to 1950 is of little value for predicting crop responses on eroded soils today. Technology advancements have masked soil productivity declines due to erosion. Although there is practically no quantification available, it is logical to expect significant crop and landscape effects. Bennett (1939) and Baver's (1950) classical reviews adequately describe the gross effects of soil erosion on crop yield during this early period.

Since 1950, only two research methods have been used extensively to measure the effects of soil erosion on soil productivity. The most frequently used technique is the cut and fill method that followed extensive land forming processes in the United States since 1950. The other approach is multiple regression analyses applied to random samples of associated crop yield and measured soil erosion. Much of this data is available only in dissertation form.[3,4] These approaches are less than desirable to assess adequately the effects of soil erosion on crop yields.

SOIL EROSION-PRODUCTIVITY ASSOCIATED WITH SOIL CLASSIFICATION

A summary of the effects of crop yield reduction due to soil erosion is shown by soil series, texture, and family classification in Tables 2 and 3. These estimates assume that adequate levels of plant nutrients and water were supplied for optimum plant growth. Further, all of these soils have histories of erosion problems by natural rainfall. They generally separate with respect to soil productivity as follows: Mollisols and Alfisols vs. Ultisols (Soil Survey Staff, AH436, 1975).

Except for the Grenada soil of west Tennessee, which is a fragipan soil, it is safe to say the initial crop yield reductions on exposed subsoils of Mollisols-Alfisols are ≤50% those of the Ultisols. These reductions on the former vary from 5 to 30%. Exceptions among these soils may occur. Some early evidence suggests the Shelby loam and associated soils may be the most erosive of the Mollisols (Bennett, 1939; Smith et al., 1945). Smith et al. (1945) reported small grain and corn yield reductions of 41 and 52%, respectively on a Shelby loam. No recent data, with high fertilization and improved germplasm, appears available to confirm the effects of soil erosion on the Shelby loam. Equivalent yield reductions on the Ultisols vary from 22 to 47%.

Because cultivated Ultisols of the southeastern USA are thinner and suffer problems associated with subsoil acidity, there is more recent research on their productivity related to soil erosion. Langdale et al. (1979b) used a watershed study to demonstrate the seriousness of a few centimeters of soil erosion on an Ultisol in the Southern Piedmont. At current nonirrigated corn production levels, each centimeter of eroded topsoil costs the producer 150 kg/ha per year (2.34 bu/acre per year) of corn grain. Soil deposition (local alluvium) did not significantly improve grain yields on the watershed. Some of these yields as well as a few reported by Adams (1949), Batchelder and Jones (1972), and Buntley and Bell (1976) are summarized in Table 4a to illustrate the expected yields of variable eroded Ultisols and a problem Alfisol. These data suggest that grain crops (corn, soybeans, and wheat) are more drastically affected by erosion than are cotton and tall fescue (*Festuca arundinacea* L.). Adams (1949) suggests that vetch (*Vicia suhu* L.) growth response exceeded that of grain and cotton crops on low fertility eroded plots. Baver (1950) also

Table 2. Family classification of erodible soils with crop data following surface soil removal.

Soil series	Surface texture	Subsoil texture	Family classification
	Deep medium textured soil		
Beadle	Silty clay loam	Silty clay	Fine, montmorillonitic, mesic, Typic Arguistolls
Chama	Silt loam	Silt loam	Fine-silty, mixed, frigid, Typic Haploborolls
Gardena	Fine sandy loam	Silt loam	Coarse-silty, frigid, mixed, Pachic Udic Haploborolls
Ida	Silt loam	Silty loam	Fine-silty, mixed (Calcareous), mesic, Typic Udorthents
Marshall	Silt loam	Silty clay loam	Fine-silty, mixed, mesic, Typic Hapludolls
Monona	Silt loam	Silt loam	Fine-silty, mixed, mesic, Typic Hapludolls
Grenada	Silt loam	Silt loam	Fine-silty, mixed, thermic, Glossic Fragiudalfs
Memphis	Silt loam	Silt loam	Fine-silty, mixed, thermic, Typic Hapludalfs
	Shallow medium to coarse textured soils		
Brandon	Silt loam	Silty clay	Fine-silty, mixed, thermic, Typic Hapludults
Cecil	Sandy clay	Clay	Clayey, kaolinitic, thermic, Typic Hapludults
Groseclose	Clay loam	Clay	Clayey, mixed, mesic, Typic Hapludults

Table 3. Estimated percent crop yield† reduction following topsoil removal.

Soil series	Corn	Cotton	Soybeans	Small grains	Forages
	Deep medium textured soils				
Beadle‡ (Olson, 1977)	17	--	--	--	--
Chama‡ (Black, 1968)	--	--	--	14	--
Gardena‡ (Carlson et al., 1961)	19	--	--	--	--
Ida‡ (Spomer et al., 1973)	8 to 30	--	--	--	--
Marshall (Engelstad et al., 1961)	13 to 17	--	--	--	--
Monona‡ (Spomer et al., 1973)	8 to 30	--	--	--	--
Grenada (Buntley et al., 1976)	26	20	40	24	17
Memphis (Buntley et al., 1976)	9	12	20	11	5
	Shallow medium to coarse textured soils				
Brandon (Buntley et al., 1976)	44	35	47	22	25
Cecil (Adam, 1949; Langdale, 1979)	40	38	22 to 31§	34	22
Groseclose‡ (Batchelder & Jones, 1972)	36	--	--	--	--

† Assumed plant nutrients were supplied in sufficient quantity to eliminate nutrient stress for the surface horizons.
‡ Studies associated with land forming disturbances.
§ Minimum (inrow chisel) and conventional tillage, respectively.

discussed some research on North Carolina soils (Ultisols) that suggests a simple shift from cultivated row crops to forage legumes masks the effects of soil erosion. However, the tolerance specificity of agronomic plant species to acid soils is well known today (Foy, 1974). A crop yield survey by Fenton et al. (1971) in Iowa shows that only the most erosive Mollisols and Entisols possess similar yield reduction trends as the Ultisols (Table 4b).

Table 4a. Crop yield estimates associated with various levels of soil erosion in—southeastern USA.

Degree of erosion	Crop yield				
	Corn	Soybeans	Cotton†	Small grains†	Forage†
	q/ha				
Memphis silt loam (Typic Hapludalfs)‡, 2 to 5% slope					
None	69	27	9.52	36	76
Eroded	65	24	9.24	35	76
Severe	60	22	8.40	32	72
Grenada silt loam (Glossic Fragiudalfs)‡, 0 to 5% slope					
None	60	27	8.40	36	72
Eroded	53	20	7.84	31	67
Severe	44	16	6.72	27	60
Brandon silt loam (Typic Hapludults)‡, 2 to 12% slope					
None	50	20	7.28	33	65
Eroded	44	13	6.72	32	60
Severe	28	11	4.76	26	49
Cecil sandy clay (Typic Hapludults)§, 2 to 10% slope					
Deposition (Local alluvium)	62	--	--	--	--
Eroded	58	21 to 31¶	13.89	24	174
Severe	19	15 to 24¶	8.66	16	137

† Buntley and Bell, 1976 (Cotton-Lint, Small Grain-Wheat, and Forage-Fescue); Adams, 1949 (Cotton-Seed, Small Grain-Oats, Forage-Vetch).
‡ Buntley and Bell, 1976.
§ Langdale et al., 1979b (Corn and Soybeans) and Adams, 1949 (Cotton, Small Grain, and Forage).
¶ Conventional and minimum (Inrow chisel 23 cm deep) tillage, respectively.

Table 4b. Crop yield estimates associated with various levels of soil erosion in—midwestern USA.

Degree of erosion	Crop yield†			
	Corn	Soybeans	Small grain	Forage
	q/ha			
Seymour silt loam (Aquic Argiudolls) 2.5 to 6.0% slope				
None	--	--	--	--
Slight	52	22	16	78
Severe	43	17	13	63
Marshall silty clay loam (Typic Hapludolls) 2.5 to 6.0% slope				
None	--	--	--	--
Slight	67	28	22	90
Moderate	62	26	20	85
Monona silt loam (Typic Hapludolls) 2.5 to 6.0% slope				
None	--	--	--	--
Slight	62	25	25	83
Moderate	56	23	23	76
Ida silt loam (Typic Udorthents) 6.0 to 9.0% slope				
None	--	--	--	--
Moderate	52	22	21	69
Severe	43	17	17	58

† Fenton et al., 1971 (Small grain—Oats, Forage—Hay).

PERMANENCY OF SOIL EROSION

Bennett (1939) and Baver (1950) both noted that productivity damage by soil erosion was persistent between 1930 to 1950. Soil management technologies such as high fertilization, improved plant germplasm, and minimum tillage with mulched surfaces were not used in studies of this era. Recently, researchers have frequently noted that crop yields on denuded soils were lower initially and improved after several years. Further, only increased quantities of N (mineral or organic) were required for corn growth to equal that on normal topsoil (Engelstad et al., 1961a, 1961b; Hays et al., 1948; Moldenhauer and Onstad, 1975; Spomer et al., 1973). Moldenhauer and Onstad (1975) used barnyard manure on 2.1-m deep construction cuts in a deep medium textured loess of western Iowa. After 2 years, their corn yield (~ 100 q/ha) on these cuts equaled those on the normal surrounding areas. They further state that barnyard manure is desirable but not necessary to restore crop yields on desurfaced loess soils of Iowa. Engelstad and Shrader (1961a, 1961b) showed that 200 kg N/ha was required initially to produce a 75-q/ha corn yield on a Marshall silt loam subsoil. Only 125 kg of N was required to obtain the same yield on an uneroded topsoil. Rosenberry's et al. (1980) energy inputs and soil erosion estimates indicate that current cost of erosion control to Iowa farmers is greater than economic return from controlling erosion. However, only short-term aggregate economic impacts were considered.

Researchers located in Montana (Ruess and Campbell, 1961) and North Dakota (Carlson et al., 1961) all showed that N as well as P and occasionally micronutrients such as Zn were deficient in soils desurfaced by land leveling. These soils were Keiser clay loam (fine-silty, mixed, mesic Ustollic Haplargids) and Gardena fine sandy loam (pachic Udic Haploborolls), respectively. Their nutrient deficiencies were correctable without great difficulty.

Batchelder and Jones (1972) found that lime, mulch, and irrigation treatment were also necessary to restore corn yields on a desurfaced Groseclose clay loam (clayey, mixed, mesic Typic Hapludults) in western Virginia. Yields on the desurfaced and normal soils of this study were not equivalent until after the 4th year. The research of Phillips and Kamprath (1973) and Thomas and Cassel (1979) involving land leveling of Atlantic Coastal Plain soils in North Carolina strongly supports the findings of Batchelder and Jones.

Soybean yields were improved with minimum tillage (inrow chisel 23 cm deep through a surface wheat mulch) on Southern Piedmont soils (Table 4)[6]. However, these yields were far below those produced by the same tillage procedure on moderately eroded soil sites. Furthermore, this tillage method appears to be of little value where saprolite is within the plow layer.

On the distant horizon, agricultural technologies appear to be gaining a little on the permanency of soil erosion. No-tillage (fluted

---

[6] Langdale, G. W. 1978. Unpublished data from the Southern Piedmont Conservation Research Center, USDA-ARS, Watkinsville, GA.

coulter) is beginning to drastically halt soil erosion and in some cases improve crop yields on lands that are highly erodible (Langdale et al., 1979a; Phillips et al., 1980).

SUMMARY AND CONCLUSIONS

The extreme but not uncommon example on a world scale where erosion reveals barren rock is so obvious as to require no documentation. On deep medium textured soils only additional N and P fertilizers and occasional micronutrients are necessary to produce crop yields on eroded soils equivalent to those of uneroded. In most other cases between the above extremes, however, the crop, the soil, and the level of technology to be applied must be specified before an accurate appraisal of the effect of erosion on crop yield can be made. In the southeast USA, loss of a few centimeters of surface soil may be very important because of thin surface topsoils and phytotoxic levels of exchangeable Al associated with acid subsoils. Some evidence in the literature indicates forage crops are more tolerant to eroded southeastern soils than grain or cotton species.

On some eroded soils, a shift from row crops to legume forages minimizes the effects of soil erosion. This is mainly due to N fixation during periods of low evapotranspiration. Minimum tillage also enhances the recovery of some of the grain crop yield potential lost to soil erosion through water conservation. In most cases, additional fertilizers, lime, mulches, and irrigation may be required to restore crop yields to a competitive economic level on eroded soils. Current resource-product price ratios associated with available technologies probably would inhibit reclamation of most severely eroded soils for row crop production. All of these factors impinge on the time period required for agronomic reclamation of eroded soils. Because of the random-spatial nature of soil erosion, randomize plot designs cannot give an accurate measure of crop yields lost on eroded soils. This research is also expensive and unrewarding to the soil scientist. To a large extent, data from cut and fill as well as land resource survey type studies were used to estimate crop yield reduction due to soil erosion on cultivated lands. These methods probably bias our estimates. Winters and Simonson (1951) stated that the scarcity of precise data bearing on the relationships of exposed subsoil properties to plant growth precludes comprehensive discussion. To a large extent this is still true almost 30 years later.

LITERATURE CITED

Adams, W. E. 1949. Loss of topsoil reduced crop yield. J. Soil Water Conserv. 4:130.
Batchelder, A. R., and J. N. Jones, Jr. 1972. Soil management factors and growth of *Zea mays* L. on topsoil and exposed subsoil. Agron. J. 64:648-652.
Baver, L. D. 1950. How serious is soil erosion. Soil Sci. Soc. Am. Proc. 42:1-5.
Bennett, H. H. 1939. Soil conservation. McGraw-Hill Book Co., Inc.
Black, A. L. 1968. Conservation bench terraces in Montana. Trans. ASAE 11:393-395.

――――, and B. W. Greb. 1968. Soil reflectance, temperature, and fallow water storage on exposed subsoils of a Brown soil. Soil Sci. Soc. Am. Proc. 32:105–109.

Black, C. A. 1968. Soil-plant relationships. John Wiley and Sons, Inc., New York.

Buntley, G. J., and F. F. Bell. 1976. Yield estimates for the major crops grown on the soils of west Tennessee. Tenn. Agric. Exp. Stn. Bull. 561. 124 p.

Carlson, C. W., D. L. Grumes, J. Alessi, and G. A. Reichman. 1961. Corn growth on Gardena surface and subsoil as affected by application of fertilizer and manure. Soil Sci. Soc. Am. Proc. 25:44–47.

Copley, T. L., L. A. Forrest, A. G. McCall, and F. C. Bell. 1944. Investigations in erosion, control and reclamation of eroded land at the Central Piedmont Conservation Experiment Station, Statesville, N.C. U.S. Dep. Agric. Tech. Bull. No. 873.

Eck, H. V., V. L. Hauser, and R. H. Ford. 1965. Fertilizer needs for restoring productivity on Pullman silty clay loam after variable degrees of soil removal. Soil Sci. Soc. Am. Proc. 29:209–213.

Engelstad, O. P., and W. D. Shrader. 1961a. The effect of surface soil thickness on corn yields: II. As determined by an experiment using normal surface soil and artificially-exposed subsoil. Soil Sci. Soc. Am. Proc. 25:497–499.

――――, O. P., W. D. Shrader, and L. C. Dumenil. 1961b. The effects of surface soil thickness on corn yields: I. As determined by a series of field experiments in farmer operated fields. Soil Sci. Soc. Am. Proc. 25:494–497.

Fenton, T. E., E. R. Duncan, W. D. Shrader, and L. C. Dumenil. 1971. Productivity levels of some Iowa soils. Spec. Rep. No. 66. Iowa Agric. Exp. Stn. 23 p.

Foy, C. D. 1974. Effects of aluminum on plant growth. p. 601–642. *In* E. W. Carson (ed.) Plant root and its environment. Univ. Press of Va., Charlottesville.

Hays, O. E., C. E. Bay, and H. H. Hull. 1948. Increasing production on an eroded loess-derived soil. J. Am. Soc. Agron. 40:1061–1069.

Heilman, M. D., and J. R. Thomas. 1961. Land leveling can adversely affect soil fertility. J. Soil and Water Conserv. 16:71–72.

Langdale, G. W., A. P. Barnett, R. A. Leonard, and W. G. Fleming. 1979a. Reduction of soil erosion by the no-till system in the Southern Piedmont. Trans. ASAE. 2:82–86, and 92.

――――, J. E. Box, Jr., R. A. Leonard, A. P. Barnett, and W. G. Fleming. 1979b. Corn yield reduction on eroded Southern Piedmont Soils. J. Soil and Water Conserv. 34:226–228.

Latham, E. E. 1940. Relative productivity of the A horizon of the acid sandy loam and the B & C horizons exposed by erosion. J. Am. Soc. Agron. 32:950–954.

Moldenhauer, W. C., and C. A. Onstad. 1975. Achieving specified soil loss levels. J. Soil and Water Conserv. 30:166–168.

Musgrave, G. W., and R. A. Norton. 1937. Soil and water conservation investigations at the Soil Conservation Experiment Station, Missouri Valley Loess Region, Clarinda, Iowa. U.S. Dep. Agr. Tech. Bull. 558.

Olson, T. C. 1977. Restoring the productivity of a glacial till soil after topsoil removal. J. Soil and Water Conserv. 32:130–132.

Parks, C. L., H. F. Perkins, and J. T. Mays. 1967. A greenhouse study of P and K requirements for ladino clover establishment on Kaolin strip mine spoil. Georgia Agr. Res. 9:8–10.

Peperzak, P., W. D. Shrader, and O. Kempthorne. 1959. Correlation of selected soil indices with plant growth on highway backslopes in Iowa. Highway Res. Board Proc. 38:622–637.

Phillips, J. A., and E. J. Kamprath. 1973. Soil fertility problems associated with land-forming in the Coastal Plain. J. Soil and Water Conserv. 28:69–73.

Phillips, R. E., R. L. Blevins, G. W. Thomas, W. W. Frye, and S. H. Phillips. 1980. No-tillage agriculture. Sci. 208:1108–1113.

Reuss, J. O., and R. E. Campbell. 1961. Restoring productivity to leveled land. Soil Sci. Soc. Am. Proc. 25:302–304.

Richardson, E. C., and E. G. Diseker. 1961. Control of roadbank erosion in the southern Piedmont. Agron. J. 53:292-294.

Rosenberry, P., R. Knutson, and L. Harmon. 1980. Predicting the effects of soil depletion from erosion. J. Soil Water Conserv. 35:131-134.

Shrader, W. D., H. P. Johnson, and J. F. Timmons. 1963. Applying erosion control principals. J. Soil Water Conserv. 18:195-199.

Smith, D. D., D. M. Whitt, A. W. Zingg, A. G. McCall, and F. G. Bell. 1945. Investigations in erosion control and reclamation of eroded Shelby and related soils at the Conservation Experiment Station, Bethany, Missouri. USDA Tech. Bull. 883.

Soil Survey Staff. 1975. Soil taxonomy. USDA, Soil Conserv. Ser. AH436. 754 p.

Spomer, R., W. D. Shrader, P. Rosenberry, and E. L. Miller. 1973. Level terraces with stabilized backslopes on loessial cropland in the Missouri Valley—A cost-effectiveness study. J. Soil Water Conserv. 28:127-131.

Thomas, D. J., and D. K. Cassel. 1979. Land forming Atlantic Coastal Plain soils: Crop yield relationships to soil physical and chemical properties. J. Soil Water Conserv. 30:20-24.

Thomas, J. R., M. D. Heilman, and L. Lyles. 1974. Predicting nitrogen fertilizer requirements after land leveling. Agron. J. 66:371-374.

Whitney, R. S., R. Gardner, and D. W. Robertson. 1950. The effectiveness of manure and commercial fertilizers in restoring the productivity of subsoils exposed by leveling. Agron. J. 42:239-245.

Winters, E., and R. W. Simonson. 1951. The subsoil. Adv. Agron. 3:1-92.

Chapter 5

Some Soil Erosion Effects on Forest Soil Productivity

G. O. KLOCK[1]

ABSTRACT

The many factors influencing the sensitivty of forest and range soils to erosion have been extensively studied. These studies have not, however, produced an accepted method to evaluate potential erosion effects on forest site productivity. A bioassay technique is suggested and its use demonstrated to make this evaluation. The general expectation is that forest soils are intolerant to erosion. Although the bioassay technique does not provide a soil loss tolerance T-value for productivity, it does show some forest soils may be less sensitive to nutrient loss through erosion.

INTRODUCTION

Degradation of mountains and soil materials accumulation in river valleys and seashore estuaries has been an active process of land form development for 3 to 4 billion years. Long before the influence of man, this erosional process formed fertile river valleys for agricultural enterprise and estuary wetlands that provide critical support for the life of the sea. Forest managers have some control within practical means over this geological erosional process, but our most fruitful management focuses on soil erosion produced by man's activities. Any increase in erosion over that of the geologic norm caused by man's activities is accelerated erosion. Forest and range management should have a goal to prevent accelerated

---

[1] Principal research soil scientist, USDA Forest Service, Pacific Northwest Forest and Range Experiment Station, Wenatchee, WA.

Copyright © 1982 ASA, SSSA, 677 South Segoe Road, Madison, WI 53711. *Determinants of Soil Loss Tolerance.*

erosion and to reduce the adverse impacts of geologic erosion. Even with this idealistic goal, accelerated erosion does occur and research is continually needed to assess erosion impacts and develop better methods of control and reduction. Economics and national policies undoubtedly affect the amount of erosion tolerated as illustrated in Timmon's and Amos's chapter regarding agricultural cropland.

The objectives of this presentation are (1) to review the factors that contribute to soil erosion on forest and associated lands and (2) to discuss an example of a method for evaluating the effects of erosion on forest site productivity.

Factors Affecting Soil Erosion in Forests

Some of the earliest forestry research in this country was directed towards reducing and preventing erosion. Early erosion research concentrated on identifying soil properties influencing erosion, studying the protective effects of vegetation, and the measurement of erosion potential using rainfall simulation.

Investigations into the influence of forest and range soil properties affecting erosion paralleled erosion research started earlier on agricultural lands. Forest soil properties generally found to relate to soil erosion were texture, porosity, organic matter content, bulk density, moisture retention characteristics, pH, and aggregation (Olson, 1949; Anderson, 1951 and 1954; Wallis and Willen, 1963; Willen, 1965; Wooldridge, 1964 and 1965; and Balci, 1968).

In early research on the influence of vegetative cover on erosion, Anderson (1951) showed forest vegetative cover density to have a highly significant effect. Chapline (1929) reported that when the vegetative cover on rangelands was increased from 16 to 40%, soil loss by erosion was reduced 56%.

Willen (1965) showed that soil erodibility also varied between cover types. Soils developed under grass were potentially the most erodible and those under conifers the least. These differences in erodibility among cover types may be a reflection of vegetative cover effect on soil genesis caused by differences in the chemical composition, pH, and amount of annual litter fall. [Kittredge (1948) and Lutz and Chandler (1946) first noted these differences in the chemical composition of forest vegetation and their possible effects on the soil erosion process.]

If the rainfall intensity exceeds the soil infiltration rate for any duration, overland flow will occur and the energy of the flowing water can cause erosion. A number of investigators have attempted to study this erosion process as well as the influence of the raindrop on soil particle movement on forest and associated rangelands. Dortignac (1951), who designed the Rocky Mountain infiltrometer, and Packer (1957) who developed an improved model known as the Intermountain Infiltrometer, were some of the earlier investigators.

Research, since the work of these early investigators, has continued to improve our techniques in identifying site factors as well as specific soil and vegetative properties which influence the erosion process (Johnson and Beschta, 1980). Researchers agree that the forest environment is generally stable with relatively minimal soil loss by erosion until severely disturbed. Nearly all accelerated erosion on forest lands follows a major perturbation such as the construction of logging roads (Patric and Brink, 1977) which increases the land exposed to the erosive forces of water or wind.

Logging roads are generally recognized as a point source of soil erosion. However, a more general example of a perturbation that can lead to severe non-point source of erosion on forest and associated rangelands is wildfire. Two excellent examples of serious erosion following wildfire are those described in southern California by Krammes (1960) and the 1970 wildfires in north central Washington State.

Forest Disturbances that may Cause Serious Erosion

In north central Washington the extensive forest land affected by wildfire provides a good example of forest perturbation and consequent erosion. Within the nearly 40,000-ha area affected by the wildfire lies the Entiat Experimental Forest. The Entiat Experimental Forest was formed from three near 500-ha watersheds covered by Douglas fir [*Pseudotsuga menziesii* (Mirb.) Franco] and ponderosa pine (*Pinus ponderosa* Laws.). For 12 years prior to the fire, sedimentation monitoring on these watersheds showed erosion to be virtually nonexistent (Helvey et al., 1976). Following fire, however, the tremendous volumes of sediment discharged from the watersheds in the forms of debris avalanches, flows, bedload, and suspended sediment were greater than that which could be effectively measured (Klock and Helvey, 1976). Obviously this was an epoch event over which land managers had little control other than the spread of lightning-caused wildfire.

Man may have the opportunity to influence the "magnitude" of epoch erosional events initiated by wildfire. On the experimental watershed and on an adjacent forest watershed equally consumed by wildfire in 1970, salvage logging proceeded following the fire and before the rainstorms that led to the extreme sedimentation in 1972. Since research has identified soil disturbance and compaction as key contributors to forest soil erosion (Sidle, 1980), considerable attention was given on the experimental watershed to protecting the forest floor using advanced logging techniques (Klock, 1975) which minimized severe soil surface disturbance. Helicopter yarding and tractor skidding over snow were used extensively in this area. On the adjacent watershed, less attention was given to surface protection on steep slopes. Traditional methods of logging were used including cable ground skidding and large tractors. Our studies showed that the advanced logging methods (mostly helicopter and skyline) generally kept soil disturbance below 25% of the total area while it was near 75% in the traditional tractor- and cable-skidded area. Following the 1972 rainstorms, evidence of surface erosion on the helicopter-logged areas of the experimental watershed was found on 3.4% of the area and 30.7 to 41.1% of the total logged area on the tractor- and cable-logged adjacent watershed (Klock, 1975). The severity of the erosion in the cable-skidded area was extreme. Although the erosion process created deep rills in some areas and left small areas relatively undisturbed, it was estimated that an average 3.8 cm (1.5 in) of surface soil was lost from the slopes (Klock, 1976). Along with the additional 12% of the area disturbed by roads to accommodate cable skidding, the erosional perturbation impact was severe both in terms of effect on future site productivity and downstream values. Obviously future wildfire salvage logging activities in the north central Washington Entiat area will require advanced logging techniques to prevent serious erosion.

The serious soil surface erosion observed on the wildfire and salvage logging affected area raised the question of what effect this event may have had on the site's future productivity. To answer this question, the author needed a quick method of assessing soil loss by erosion on site productivity as the ultimate site effect may not be known for a rotational period, perhaps 100 years. A survey of available literature showed that a bioassay technique described by Meurisse and Youngberg[2] may be useful in making site productivity assessments.

Productivity

The 1976 National Forest Management Act directs forest and range managers to carry out their management activities such that they do not reduce or impair the future productivity of the land. Soil erosion is a process that can be accelerated by negligent management such that future site productivity may be reduced.

Site Productivity Indexing

Site productivity on forest and rangelands is generally thought of as a measure of the ability of a site to produce timber, forage, wildlife, or other biological outputs. It is generally assumed that there is no single measurement that adequately describes the productivity of forest and range sites for all of the products or outputs that can be obtained from it. Most often there are only measurements in terms of volume or board length of timber (site index) or weight of forage produced per acre annually as estimates of site productivity.

While this definition of site productivity may illustrate the total potential productivity of a site, it appears appropriate to define site productivity in terms of a site's ability to sustain net primary production (npp). Net primary production is the differential output of site biomass

---

[2]Bioassay of four ando-like soils from a western hemlock forest in Oregon. Manuscript under preparation.

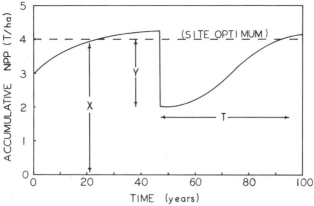

Fig. 1. Influence of forest and range successional dynamics on net primary production (npp). The total potential site productivity is shown as X, effect of site perturbation as Y, and T as the time period of restoration following forest or rangeland perturbation.

less that biomass converted to detritus and that which feeds consumers and can be related to the value of forest products such as board length per unit of timber. In Fig. 1, for purpose of discussion, the influence of forest and range successional dynamics on npp is illustrated. There are three important parameters within this relationship. First, potential site productivity (X) needs to be estimated. Site perturbations (Y), although illustrated in Fig. 1 as a negative effect on npp, can in many cases also have a positive effect. Forest fertilization may be one such case. The most important factor is time (T). Since natural ecosystems are rather resilient, most sites will eventually return to their potential level of productivity (X) given enough time. Under intensive levels of forest and range management, however, we may not be able to afford the time span required for nature to heal some forms or magnitudes of perturbations.

A number of factors determine a site's productivity or its npp output although two factors—climate and soil—predominantly have the greatest influence. If we assume climate remains constant and that we are not manipulating vegetative characteristics through genetics, etc., soil factors affecting the vegetative growth characteristics will have the most important influence on site productivity. Therefore, erosion that removes the nutrient-rich surface soil can have a highly influential impact on npp. Although some of this productivity can be restored by energy consuming inorganic fertilization, the time requirements in the factors of soil formation described by Jenny (1941) become alarmingly important. Essentially no information is presently available on the magnitude of site productivity reduction or extension of time required to return the site to optimum yield conditions following serious erosion on forest lands. There appears to be an opportunity with careful experimental control to examine the potential effects on erosion as well as other perturbations of the forest and range environment on site productivity through the use of bioassay analyses.

Bioassay Analyses

Greenhouse bioassay analyses have long been used in plant nutrition and fertility studies to determine fertilization response of agricultural soils. Several researchers have studied forest nutrition problems with this method (Mead and Pritchett, 1971; Swan, 1960 and 1966; Tam, 1964; Youngberg and Dyrness, 1965; Waring and Youngberg, 1972; and Meurisse and Youngberg[2]). They point out that the most obvious advantage of greenhouse bioassays on forest soils is that experimental results are obtained in a much shorter time than field studies. They also point out the inherent difficulties in extrapolating results beyond the conditions imposed in the greenhouse to forest conditions. Up to this time, no attempt has been made to determine the effects of erosion on the plant nutritional relationships of forest and range soils through the use of bioassay analyses.

Recently, the author studied the possible relationship of forest soil loss by erosion to site productivity through the use of a soil bioassay analysis. Results of this study indicate that the method developed may be a useful technique in assessing the effects of erosion on site productivity decline.

Study Method

The greenhouse bioassay analysis technique was used experimentally on four north central Washington forest and range soils to test the potential effect of surface mineral soil loss by erosion on biomass productivity. One soil developed from extrusive basalt (mollisol) and three soils developed on volcanic tephra (inceptisols) were chosen for the study. All four soils are relatively young with profile development only weakly evident. Forest floor organic material above the mineral soil was highly variable and ranged from 1 to about 7 cm deep. Conifer root distributions observed on all four soils indicate that plant nutrition requirements are generally met within the upper 30 cm (12 in) of the soil profile. Thus, the author chose his total sampling depth as the upper 30 cm of the profile. To simulate erosion, surface layers of 3, 7.5, and 15 cm (1, 3, and 6 in) were removed successively while field sampling. Soil samples were carefully removed with a spade so that equal volumes of the entire sampling depth were collected. The soil sample was placed in a small container, thoroughly mixed, and the volume needed for the bioassay analyses (normally about 10 kg) was returned to the greenhouse to be air-dried. The four sampling depths, 0 to 30, 3 to 30, 7.5 to 30, and 15 to 30 cm (0 to 12, 1 to 12, 3 to 12, and 6 to 12 in), were collected at six random sampling sites or replications within a 15-m (46.5-ft) diam area. Serious erosion is often observed on forest soils where the organic surface layers have been damaged or destroyed. Thus, the entire forest floor organic material above the mineral soil was carefully removed prior to sampling at each site. If the layer had been included in the mineral soil, the bioassays would have been confounded.

Following sieving with an 8-mm (0 to 0.2 in) screen to remove large roots and debris, 1.3 kg (2.9 lbs) of the air-dried soil from each sample were mixed with 130 ml of perlite and placed in a 15-cm (6-in) diam pot. The inert perlite was used to prevent compaction of the potting soil and to maintain adequate infiltration characteristics during the period of analyses. Two pots were filled with a sample from each location, replication and depth increment. One pot remained unfertilized and the second was fertilized at the rate of 100 kg/ha (90 lbs./acre) N, P, and K, and 17 kg/ha (15.3 lbs./acre) S. Thus, eight pots—four fertilized and four unfertilized—in a randomized block experimental design with six replications were filled from each sampling site for a total of 48 test pots from each soil for each test plant species used. All pots were saturated with distilled water from the base of the pot and allowed to drain before planting.

The bioassay analyses were run using three conifer species and one grass species.

Ten orchardgrass (*Dactylis glomerata* L.) seeds each were planted in 15-cm (6-in) diameter pots filled with soil from each location, replication, and depth increment, then thinned to five plants after germination. All pots were watered as necessary with distilled water. Excessive watering was discouraged to minimize loss of nutrients by drainage. After 60 days growth, the biomass accumulated above the pot's soil surface was cut, ovendried, and weighed.

Ponderosa pine, Douglas fir, and lodgepole pine (*Pinus contorta* Dougl.) seed collected in the same region as the soils sampled were germinated in perlite within a controlled atmosphere chamber. Following emergence five small seedlings were transplanted to each test pot. Root length at this time was generally less than 7 cm (3 in). Again pots were irrigated as necessary with distilled water. All conifer seedlings were grown for 6 months following transplanting under a 16-hour photoperiod within the greenhouse. At the end of the test period, the seedlings were cut at the pot surface, ovendried at 60 C for 48 hours, and weighed to determine biomass accumulation. An analysis of variance was used to test differences in biomass accumulation among sample depth increments.

Previous studies (Klock et al., 1975) have shown that N is the most limiting plant nutrient, with S possibly of secondary importance for plant growth on north central Washington forest and range soils.

Results

Average dry weight biomass produced per pot and relative yields expressed as a percent of the total unfertilized yield for each species on the 0 to 30-cm (0 to 12-in) sampling increment for the four test soils are shown in Table 1. These relative yields are illustrated for visual comparison in Fig. 2 through 5.

Table 1 and Fig. 2 show the dramatic effect of removing the upper 3 cm (1 in) of surface soil from the Brennegan watershed sampling location on biomass production. Soils in the Brennegan watershed have developed

Table 1. Dry weight biomass yields from four soil sample depths on forest soils from four different areas of north central Washington. Plants for each soil type were grown under greenhouse conditions during different time periods.

Species (fert.)†	0–30 cm yield/gm/pot	%‡	3–30 cm yield/gm/pot	%‡	7.5–30 cm yield/gm/pot	%‡	15–30 cm yield/gm/pot	%‡
Brennegan			Brennegan Soil					
Orchardgrass (U)	1.23 ± 0.22	100	0.16 ± 0.08	13	0.12 ± 0.11	9	0.17 ± 0.12	14
Orchardgrass (F)	9.04 ± 0.91	735	9.07 ± 1.31	737	9.19 ± 0.71	747	8.21 ± 0.67	668
Ponderosa pine (U)	3.81 ± 0.68	100	0.55 ± 0.18	14	0.63 ± 0.25	17	0.51 ± 0.39	13
Ponderosa pine (F)	4.11 ± 0.89	126	4.15 ± 0.97	109	4.12 ± 2.08	108	4.25 ± 0.48	111
Douglas fir (U)	1.95 ± 0.95	100	0.29 ± 0.07	15	0.27 ± 0.05	14	0.26 ± 0.04	14
Douglas fir (F)	3.89 ± 2.07	200	3.18 ± 1.40	163	4.30 ± 1.86	220	3.09 ± 1.34	158
Lodgepole pine (U)	1.09 ± 0.44	100	0.16 ± 0.06	15	0.14 ± 0.03	13	0.18 ± 0.06	16
Lodgepole pine (F)	1.91 ± 0.90	175	2.07 ± 0.45	189	1.82 ± 0.71	167	2.07 ± 0.74	190
			Shady Pass Soil					
Orchardgrass (U)	0.58 ± 0.11	100	0.31 ± 0.05	53	0.27 ± 0.03	46	0.24 ± 0.07	42
Orchardgrass (F)	7.02 ± 0.88	1,220	6.62 ± 1.31	1,141	7.00 ± 1.19	1,207	8.28 ± 1.01	1,427
Ponderosa pine (U)	0.51 ± 0.06	100	0.53 ± 0.05	105	0.54 ± 0.05	105	0.51 ± 0.05	100
Douglas fir (U)	0.30 ± 0.21	100	0.24 ± 0.02	81	0.22 ± 0.05	74	0.24 ± 0.03	81
			Colockum Soil					
Orchardgrass (U)	2.33 ± 0.71	100	1.89 ± 0.37	81	1.51 ± 0.24	65	1.17 ± 0.21	50
Orchardgrass (F)	10.11 ± 1.39	434	9.40 ± 0.84	403	9.88 ± 1.28	424	9.19 ± 0.90	394
Ponderosa pine (U)	0.49 ± 0.14	100	0.38 ± 0.06	78	0.42 ± 0.07	86	0.31 ± 0.04	63
Ponderosa pine (F)	1.64 ± 0.38	335	1.48 ± 0.42	302	1.33 ± 0.27	271	1.47 ± 0.22	300
Douglas-fir (U)	0.78 ± 0.07	100	0.58 ± 0.11	75	0.61 ± 0.12	78	0.65 ± 0.15	84
Douglas-fir (F)	1.90 ± 1.20	244	1.92 ± 0.72	247	2.51 ± 0.64	321	1.66 ± 0.62	213
Lodgepole pine (U)	2.06 ± 0.19	100	1.43 ± 0.31	69	1.21 ± 0.17	59	1.24 ± 0.15	60
Lodgepole pine (F)	7.89 ± 0.87	383	6.32 ± 0.77	306	5.54 ± 1.38	269	5.70 ± 1.11	277
			Buck Meadows Soil					
Orchardgrass (U)	2.01 ± 0.50	100	0.46 ± 0.23	23	0.53 ± 0.20	26	0.14 ± 0.04	7
Orchardgrass (F)	13.50 ± 0.88	675	12.80 ± 1.55	640	14.33 ± 1.37	727	12.29 ± 0.81	615
Ponderosa pine (U)	4.82 ± 1.78	100	2.21 ± 0.52	46	2.60 ± 0.19	54	1.00 ± 0.66	21
Douglas fir (U)	2.90 ± 0.58	100	1.05 ± 0.20	36	1.27 ± 0.34	44	0.68 ± 0.10	23

† (U) signifies unfertilized and (F) fertilized.
‡ Relative yield determined by (yield per pot/yield per pot from unfertilized 0 to 30 cm soil sample) × 100.

EROSION EFFECTS ON FOREST SOILS 61

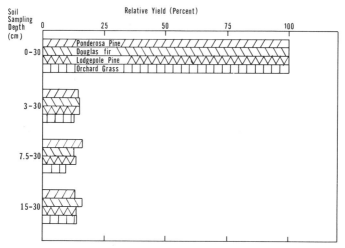

Fig. 2. Average relative yields of biomass produced by four plant species as affected by successive removal of surface layers of soil from the Brennegan site with and without fertilization. For all species 100% relative yield is equivalent to the biomass produced on the 0 to 30 cm depth soil sample.

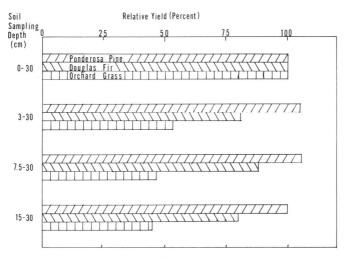

Fig. 3. Average relative yields of biomass produced by four plant species as affected by successive removal of surface layers of soil from the Shady Pass site with and without fertilization. For all species 100% relative yield is equivalent to the biomass produced on the 0 to 30 cm depth soil sample.

on volcanic tephra from Glacier Peak nearly 35 km (21 m) to the northwest. The sampling location was under openly stocked intermediate age (near 100-year) ponderosa pine. The site had been burned by ground fire 4 years before sampling. Although erosion depths greater than 3 cm (1 in) were reported in this watershed prior to our sampling date (Klock, 1976), there was no evidence of surface erosion at our sampling sites on a 20%

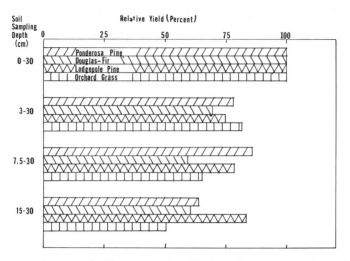

Fig. 4. Average relative yields of biomass produced by four plant species as affected by successive removal of surface layers of soil from the Colockum site with and without fertilization. For all species 100% relative yield is equivalent to the biomass produced on the 0 to 30 cm depth soil sample.

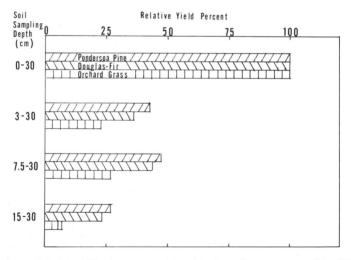

Fig. 5. Average relative yields of biomass produced by four plant species as affected by successive removal of surface layers of soil from the Buck Meadows site with and without fertilization. For all species 100% relative yield is equivalent to the biomass produced on the 0 to 30 cm depth soil sample.

slope. No erosion was present as the area was lightly burned and protected from salvage logging. The response of the conifer seedlings as well as the orchardgrass to fertilization at all sample depth increments appears to indicate that the major bioassay response is due to different fertility levels. Results therefore indicate that the soil fertility is most critical in the uppermost 3 cm of the soil profile in the Brennegan watershed.

Although the Shady Pass sampling location is on soils developed on the same Glacier Peak volcanic tephra as soils in the Brennegan watershed, the plant growth response with bioassay analyses is quite different (Table 1 and Fig. 3). Removing surface soil had a marked difference in bioassay yields only for the orchardgrass. In fact, soil surface removal had little influence on the biomass accumulations of ponderosa pine. Lodgepole pine was not used as a test species, and the ponderosa pine and Douglas fir were not tested with fertilization on soils from Shady Pass. The orchardgrass responded extremely well to fertilization or near twice that for the Brennegan watershed soil. The 6-month conifer biomass accumulation on the Shady Pass soil samples was only 25% of the growth of the Brennegan watershed soil. These soils were not tested with the bioassay analysis at the same time so a statistical analysis of this difference is not possible. The greenhouse conditions (temperature, light, etc.) were maintained identically during the test for both soils. The Shady Pass sampling site was under a well stocked stand of large old-growth ponderosa pine. These bioassay results might indicate that a large percentage of plant available N is withheld in the forest floor or the biomass presently occupying the Shady Pass site.

Soils tested from the Colockum location have developed predominantly from Columbia River basalt although there is evidence of volcanic ash in the surface soil. The sampling site supports a rather young (less than 50-year) low elevation stand of ponderosa pine. The area is heavily used for grazing both by cattle and big game including elk. Potential loss of future growth by erosion on this soil appears to be moderate as indicated by bioassay analyses (Table 1 and Fig. 4). However, again the most sensitive increment to erosion appears to be the upper 3 cm (1 in) of soil profile. The depth increment sampling appeared to have little influence on the response to fertilization for all test species.

The Buck Meadow sampling site was another location where removal of the upper 3 cm (1 in) of surface mineral soil had a marked effect on bioassay yields and potential site productivity loss with soil surface removal by erosion (Table 1 and Fig. 5). As was the case with the Shady Pass soil evaluation, lodgepole pine was not tested and the conifers were not fertilized in our study. The soils from the Buck Meadow sampling site were developed from volcanic ash, possibly from Mt. Rainier, and support a mature mixed stand of conifers. Of the four soils tested, the forest floor organic matter was deepest at this location. Fine conifer roots were observed within this material. Again these results might indicate that our test results reflect the distribution of plant available N as influenced by nutrient cycling. It apepars that most of the site N requirements are met by the distribution of available N in the forest floor and the upper 3 cm (1 in) of mineral soil. Therefore, any land management activity such as the erosion which destroys the forest floor and displaces the upper 3 cm (1 in) of surface soil could have a marked negative impact on the future productivity of this site. This site is under further study to determine the impacts of alternative forest management strategies.

Some important general observations were made during the course of our bioassay analyses study. Waring and Youngberg (1972) pointed out

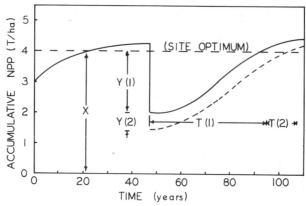

Fig. 6. Potential effect of erosion on net primary production (npp) following perturbation. The total potential site productivity (X) is little affected by erosion. The additional effects of erosion on productivity (Y_2) by a forest and range perturbation (Y_1), as possibly measured by bioassays, could be evaluated in terms of the additional time (T_2) required for optimum production restoration.

that plant productivity is a function not only of soil fertility and plant nutrition, but also of the genetic potential of a plant and limitations upon that potential from the entire environment. Regardless of these important constraints, it appears that the study method reported here reflects true plant nutrition availability and does not reflect other artifacts created such as soil compaction, water relations, etc. Therefore, the absolute values of biomass produced are somewhat inconsequential, but the relative differences in biomass accumulation among sampling depth increments appear realistic. The orchardgrass appears to be more sensitive to fertility levels than conifers as reflected by biomass production levels. Of the conifer test species, the pine appears to be the least sensitive to soil fertility conditions with Douglas fir being somewhat intermediate. These results tend to support field observations where pine appears capable of supporting itself under lower fertility regimes than fir. These observations may lend credit to lodgepole pine being an effective re-establishment conifer species following serious soil disturbance on low fertility interior forest lands.

Although the biomass accumulation levels were variable among diffeent test plant species, with the exception of the orchardgrass bioassay on the soil from the Shady Pass site, all species showed basically the same magnitude of treatment response to surface soil removal.

Study Implications

Even though the potentials for complete surface soil erosion loss on forest and range soils of 3 to 15 cm (1 to 6 in) are extremely remote, even under the poorest levels of land management, our bioassay analyses results reflect the plant nutritional sensitivity of surface soil. These study results obviously indicate that the surface soil as well as the forest floor organic matter must be protected on juvenile forest soils to maintain or enhance future site productivity as suggested by the National Forest Management Act of 1976.

In the study reported here only the relative potential impacts on site productivity by the loss of surface soil in forested areas through the use of greenhouse bioassays are discussed. Predicting the magnitude of the erosional effect on future site productivity or biomass production from experimental bioassays, however, is difficult if not impossible under present knowledge of forest ecosystems. Under stable climatic conditions, the present level of site productivity of forest and range ecosystems can be restored by natural means regardless of the serious immediate productivity loss by accelerated erosion, as long as erosion is controlled in the future. The real management concern is that the time period necessary for productivity restoration is dramatically increased by greater levels of erosion. As pointed out earlier, at some point the time period of productivity restorations may be greater than we can tolerate or afford. Using the assumption provided to develop Fig. 1, erosion effects on both the immediate productivity level and the time of full productivity restoration by natural means are demonstrated in Fig. 6.

Bioassay analyses can be used to interpret the relative seriousness of surface soil removal by erosion on site productivity under different environmental conditions. In evaluating the long-term effects of erosion on future growth, the use of prognosis growth models as illustrated in Fig. 1 and 6 along with bioassays can be helpful. We cannot fully predict the magnitude of the erosion effect on future productivity, however, as more information is needed than can be obtained by soil sampling and the use of bioassays by themselves. For example, it is essential to know the distribution of plant nutrients by compartments and transfer rates between these compartments including the soil before anyone can effectively predict a possible tolerance level for accelerated soil erosion on forest and associated range soils.

Bioassays do not provide soil loss tolerance T-values for erosion assessment by themselves. Since soil erosion is extremely rare when the profile is protected by the natural forest floor, concern for T-values are only important when the forest floor has been removed. Bioassays do point out that there is a relative difference in magnitude of productivity loss by different soils and that the most critical zone is the soil immediately below the forest floor. Therefore, soils showing small losses in productivity by surface soil removal as simulated in a soil loss/bioassay analysis will have a relatively high soil erosion loss tolerance.

LITERATURE CITED

Anderson, H. W. 1951. Physical characteristics of soils related to erosion. J. Soil Water Conserv. 6:129–133.

―――――. 1954. Suspended sediment discharge as related to streamflow, topography, soil and land use. Am. Geophys. Union Trans. 35(2):268–281.

Balci, A. N. 1968. Soil erosion in relation to properties of eastern and western Washington forest soils. Soil Sci. Soc. Am. Proc. 42(3):430–435.

Chapline, W. R. 1929. Erosion on range lands. J. Am. Soc. Agron. 21:423–429.
Dortignac, E. J. 1951. Design and operation of Rocky Mountain infiltrometer. USDA For. Serv. Rocky Mt. For. and Range Exp. Stn. Paper No. 5. 68 p.
Helvey, J. D., W. B. Fowler, G. O. Klock, and A. R. Tiedemann. 1976. Climate and hydrology of the Entiat Experimental Forest Watersheds under virgin forest cover. USDA For. Serv. Gen. Tech. Report PNW-42. 18 p.
Jenny, H. 1941. Factors of soil formation. McGraw-Hill Book Co., Inc., New York. 281 p.
Johnson, M. G., and R. L. Beschta. 1980. Logging, infiltration capacity and surface erodibility in western Oregon. J. For. 78(6):334–337.
Kittredge, J. 1948. Forest influences. McGraw-Hill Book Co., Inc., New York.
Klock, G. O. 1975. Impact of five postfire salvage logging systems. J. Soil Water Conserv. 30(2):78–81.
―――. 1976. Estimating two indirect logging costs caused by accelerated erosion. USDA For. Serv. Gen. Tech. Publ. PNW-47. 9 p.
―――, J. M. Geist, and A. R. Tiedemann. 1975. Response of orchard grass to sulphur-coated urea on several forest and range soils. Sulphur Inst. J. 1(3–4):4–8.
―――, and J. D. Helvey. 1976. Debris flows following wildfire in north central Washington. p. 91–98. In Proc. 3rd Fed. Inter-Agency Sed. Conf., Denver, Colo.
Krammes, J. S. 1960. Erosion from mountain side slopes after fire in southern California. USDA For. Serv. Res. Note PNW-171.
Lutz, J. F., and R. F. Chandler, Jr. 1946. Forest soils. John Wiley and Sons, Inc., New York.
Mead, D. J., and W. L. Pritchett. 1971. A comparison of tree responses to fertilizers in field and pot experiments. Soil Sci. Soc. Am. Proc. 35:346–349.
Olson, O. C. 1949. Relations between soil depth and accelerated erosion on the Wasatch Mountains. Soil Sci. 67(6):414–418.
Packer, P. E. 1957. Intermountain infiltrometer. USDA For. Serv. Intermountain For. and Range Exp. Stn. Misc. Pub. 14. 41 p.
Patric, J. H., and L. K. Brink. 1977. Soil erosion and its control in the eastern forest. p. 362–368. In Soil erosion: Prediction and control. Proc., Purdue Univ. Natl. Conf. on Soil Erosion, Spec. Publ. 21, Soil Conserv. Soc. Am., Ankeny, Iowa.
Sidle, R. C. 1980. Impacts of forest practices on surface erosion. USDA Forest Serv. Ext. Publ. PNW-195. 16 p.
Swan, H. S. D. 1960. Mineral nutrition of Canadian pulpwood species. I. The influence of nitrogen, phosphorus, potassium, and magnesium deficiencies on the growth and development of white spruce, black spruce, jack pine, and western hemlock seedlings grown in a controlled environment. Pulp and Paper Res. Inst. of Canada, Woodlands Res. Index No. 116. 66 p.
―――. 1966. Studies of the mineral nutrition of Canadian pulpwood species. The use of visual symptoms, soils and foliar analysis and soil bioassays for the determination of the fertilizer requirements of forest soils. Pulp and Paper Res. Inst. of Canada, Woodlands Res. Index No. 179. 11 p.
Tamm, C. O. 1964. Determination of nutrient requirements of forest stands. Int. Rev. For. Res. 1:115–170.
Wallis, J. R., and D. W. Willen. 1963. Variation in dispersion ratio, surface aggregation ratio and texture of some California surface soils as related to soil forming factors. Int. Assoc. Sci. Hydrol. Bull. 8(4):45–58.
Willen, D. W. 1965. Surface soil texture and potential erodibility characteristics of some Southern Sierra Nevada forest sites. Soil Sci. Soc. Am. Proc. 29:213–218.
Waring, R. H., and C. T. Youngberg. 1972. Evaluating forest sites for potential growth response of trees to fertilizer. Northwest Sci. 46(1):67–75.
Wooldridge, D. D. 1964. Effects of parent material and vegetation on properties related to soil erosion in Central Washington. Soil Sci. Soc. Am. Proc. 28(3):430–432.
―――. 1965. Soil properties related to erosion of wildland soils in Central Washington. p. 141–152. In Forest-soil relationships in North America. Oregon State Univ. Press, Corvallis.
Youngberg, C. T., and C. T. Dyrness. 1965. Biological assay of pumice soil fertility. Soil Sci. Soc. Am. Proc. 29:182–187.

Chapter 6

Determinants of Soil Loss Tolerance for Rangelands[1]

J. ROSS WIGHT AND F. H. SIDDOWAY[2]

ABSTRACT

Rangelands, which occupy about 40% of the earth's surface, have been increasingly recognized as an important and often mismanaged natural resource. Accelerated erosion is recognized as a major problem associated with the management of this vast resource. Attempts to extend the concept of soil loss tolerance (T value), as developed for cropland, to range ecosystems is of questionable validity. The fragility of rangeland ecosystems, the irreversibility of erosion damage, and the large margins of error associated with soil loss estimates make it difficult to develop T values for rangelands.

INTRODUCTION

Rangeland is defined by the Society for Range Management as land on which the native vegetation (climax or natural potential) is predominantly grasses, grass-like plants, forbs or shrubs suitable for grazing or browsing use. It includes land vegetated naturally or artificially to provide a forage cover that is managed like native vegetation. Rangelands include natural grasslands, savannahs, shrublands, most deserts, tundra, alpine communities, coastal marshes, and wet meadows (Kothmann, 1974).

---

[1] Contribution from the Northwest Watershed Research Center, ARS, USDA; Northern Plains Soil and Water Research Center, ARS, USDA; and Bureau of Land Management, USDI; in cooperation with the Agricultural Exp. Stn., Univ. of Idaho, Moscow, ID.

[2] Range scientist and soil scientist, respectively; USDA-ARS, 1175 South Orchard, Suite 116, Patti Plaza, Boise, ID 83705, and USDA-ARS, P.O. Box 1109, Sidney, MT 59270.

Of the earth's 13.8 billion ha of land surface, about 40% is classified as rangeland (Branson et al., 1972). Forest and cultivated lands account for another 30 and 10%, respectively. The 913 million ha of land in the United States are similarly distributed with about 38, 29, and 18% of the land area classified as rangeland, forest, and cultivated land, respectively (USDA, Resources Conservation Act Coordinating Committee, 1980b). These rangelands include about 345 million ha (not including forest land managed primarily for grazing) of private and federally owned land, and provide a habitat for several million wild and domestic animals. In 1976, rangeland plus grazed forest land provided about 15% of the total roughage used by U.S. livestock; enough for the annual needs of 18 million animal units (Cutler, 1980). Representing 38% of the land area, rangelands intercept a large portion of the precipitation and are important in terms of the quantity and quality of the nation's water resources.

Erosion from rangelands can be serious. These lands are usually too steep, too dry, or otherwise too nonproductive to cultivate. Langbein and Schumm (1958) generalized that, under natural vegetation such as occupies rangeland, erosion reaches a maximum with annual precipitation of 25 to 38 cm. Below 25 cm, the volume of runoff is too small to cause significant erosion; above 38 cm, precipitation supports a vegetation cover adequate to minimize or reduce erosion as compared with that from bare soil. About 99% of U.S. nonfederal rangeland and probably a similar portion of the federal rangeland is located west of the 96th meridian (USDA, Resources Conservation Act Coordinating Committee, 1980b), and a large portion of this area is located within the 25 to 38 cm annual precipitation zone. Renard (1980) stated that sediment yields from many rangeland areas in the western USA are larger than might be expected under the associated low rainfall. In addition to sparsity of vegetation, steep topography, and low infiltration, he also identified high intensity thunderstorms and steep alluvium-filled channels as causative factors.

On a per unit area basis, rangelands are relatively low in economic value and have not received the magnitude of management and research inputs as have forest and cultivated lands. Recently, two major factors have helped focus attention on the rangeland resource and its management: (1) increasing populations with increasing food requirements, and (2) increasing awareness of, and a need to improve environmental quality. Recent legislation, such as the Federal Water Pollution Control Act Amendments of 1972, the Federal Land Policy and Management Act of 1976, and the Public Rangeland Improvement Act of 1978, is indicative of the nation's aroused concern to manage for the protection of this important resource. However, in all of the planning and discussion of research and management needs, little, if anything, has been said concerning soil loss tolerances (T values) for rangelands. The purpose of this paper is to discuss T values for rangelands and problems associated with their establishment.

SOIL LOSS FROM RANGELANDS

Measurement

Soil loss is difficult to measure. Rainfall simulation and small gaged watersheds provide an opportunity to measure actual soil loss from specific sites. However, soil loss measurements from small plots are not directly applicable to large watersheds and such data are used primarily to index site, precipitation, and management effects and as inputs to predictive equations. Most soil loss data has been estimated from sediment measurements, model outputs, or prediction equations, notably the Universal Soil Loss Equation (USLE) (Wischmeier and Smith, 1978). Renard (1980) reviewed several common techniques for estimating sediment yield from rangelands. They included: (1) sediment rating-curve/flow duration method; (2) sediment delivery ratio method; (3) reservoir sediment deposition surveys; (4) field measurements of erosion and deposition; (5) bedload relationships; (6) mathematical simulation models; and (7) predictive equations. Movement of fallout ^{137}Cs has been used as an indirect measure of soil loss (Ritchie et al., 1974). Experimental watersheds, such as the USDA, ARS watersheds near Tombstone, Ariz. (15,000 ha) and near Boise, Idaho (24,300 ha), provide sediment yield data from large rangeland watersheds. The relationship between soil loss and sediment yield is discussed by Wischmeier and Smith (1978), and they indicated that the sediment delivery ratio varied approximately as the 0.2 power of the drainage area.

The soil loss data (rill, sheet, and wind erosion), reported in the recent Soil Conservation Service's (SCS) National Resources Inventory (USDA, SCS, 1978) and referenced in this paper, were calculated from the USLE and WEQ (Wind Erosion Equation) (Skidmore and Woodruff, 1968), both predictive equations. Blackburn (1980) concluded that "although 40 years of research and application have gone into development of USLE for use on cropland, research and development on the equation is not adequate for the varied conditions of western rangelands". As stated by Renard (1980), the methods for estimating sediment yields from rangelands have been developed where cultivated agriculture is prevalent and are generally not directly applicable to rangelands.

Quantity

The SCS (USDA, Resources Conservation Act Coordinating Committe, 1980b) estimated soil loss as a result of water and wind erosion from the 165 million ha of the nation's nonfederal rangeland, excluding Alaska, averages 11.6 metric tons/ha per year. Soil loss from nonfederal rangeland in Mississippi, Arkansas, Texas, New Mexico, Arizon, California, and Colorado exceeds 11.2 metric tons/ha per year; in Kansas and Wyoming, between 6.7 and 11.0 metric tons/ha per year; in Oklahoma and Nebraska, between 4.5 and 6.5 metric tons/ha per year; and in Florida, Louisiana, Missouri, Minnesota, North Dakota, South Dakota; Montana, Idaho, Nevada, Oregon, and Washington, less than 4.2 metric tons/ha per year. The remaining states have only a slight or no rangeland erosion problem. On the above hectarages, water erosion is the primary problem, averaging 7.6 metric tons/ha per year, whereas wind erosion averages 4.0 metric tons/ha per year. Only New Mexico has a severe wind erosion problem on rangeland; Texas, a moderate problem. Five other states have recognized wind erosion problems—Arizona, California, Nevada, Utah, and Washington—but no data are available which define the severity. Rangeland erosion is negligible on the small hectarage of Class I through III lands (15% of nonfederal rangelands). Class VII and VIII lands (35% of nonfederal rangelands) erode at a rate of 23.1 metric tons/ha per year. The erosion from many of these lands includes that of geologic origin and, therefore, is difficult to control within established soil loss tolerances.

The SCS (USDA, SCS, 1975) identified 42% or 68.8 million ha of rangeland in the nine western contiguous states as needing erosion control treatments.

Control

Erosion on rangelands is controlled primarily by protecting and maintaining the vegetation cover with grazing management. Steep topography and unfavorable economics limit use of cultural practices, such as seeding and fertilization. Land surface modifications, such as pitting and contour furrowing, which are designed to hold precipitation where it falls, have proved beneficial on many sites (Wight, 1976). The beneficial effects of such treatments are twofold: (1) decreased runoff; and (2) increased plant growth, which, in turn, reduces runoff. Again, topography and economics limit the application of these treatments.

DETERMINING SOIL LOSS TOLERANCE

The estimated average T value for rangelands is currently 4.5 metric tons/ha per year (USDA, Resources Conservation Act Coordinating Committee, 1980a). Criteria for establishment of rangeland T values have been essentially the same as those for cropland. The difference in the average cropland T value of 11.2 metric tons/ha per year and rangeland T value (4.5 metric tons/ha per year) is due primarily to rangelands' shallower soils. Wischmeier and Smith (1978) indicated that the cropland T values, established in 1961 and 1962, have generally proved adequate for sustaining high production, although the 16-year time lapse is perhaps a little short to draw such a conclusion. They also pointed out that T values established for sustained productivity may differ considerably from T values established for water quality control.

It is sometimes difficult to apply the T value concept for water quality control purposes, because T values do not reflect the channel erosion effects. For example, runoff from given land forms may contribute little sediment until the water becomes channelized. Channel erosion is often the major problem in terms of silting of farm ponds and reservoirs and causing other adverse downstream pollution problems. Formation of a channel may result in headcutting and progressive erosion from lower to higher elevations of the watershed. Water erosion deposits about 3.6 billion metric tons soil/year in the waterways of the 48 contiguous states (Wadleigh, 1968), a sizeable fraction of which probably comes from western rangelands.

A question of real concern is whether or not the same criteria used for establishing cropland T values can be applied to rangeland. At least two facets of the problem should be considered in answering this question: (1) the vast differences in the two ecosystems and their management, and (2) the question of geologic erosion vs. accelerated erosion on rangeland.

Cropland is land that, because of favorable soil, topography, and adequate moisture, can be cultivated and the vegetation cover controlled, at least partially, on an annual basis. If severe erosion should occur due to a periodic drought, high-intensity precipitation event, or poor management, productivity and plant cover can usually be economically restored by management. Most cropland soils are relatively deep and can sustain some soil loss without serious loss of productivity. Also, because of a more favorable soil water regime, soil formation is more rapid on cropland than on rangeland.

Rangeland is inherently more fragile than cropland. It is characterized by steep slopes, shallow soil mantles, and a native plant cover that is in delicate balance with the environment. About 71% of U.S. rangelands have slopes greater than 12%, as compared with only 10% of the cropland (USDA, Resources Conservation Act Coordinating Committee, 1980b). If a catastrophy (either natural or man-made) occurs that reduces the vegetation cover beyond some critical point, erosion may accelerate to the point that the entire soil mantle is lost (Fig. 1). Once an accelerated erosion cycle begins, it is often self-sustaining. Loss of vegetation cover through fire, drought, overgrazing, or such may result in substantial soil losses. These losses, in turn, reduce a range site's potential to produce vegetation cover and the original level of cover cannot be reestablished or maintained. Thus, as cover is reduced, soil loss increases, and as soil loss increases, cover is reduced. Also, the gullies and channels that are formed cannot readily be reshaped and stabilized like they could if they were on cultivated land.

Because of rangelands original low value per unit of land and the low probability of realizing a profit from capital investment in range improvement practices once this land is damaged, it is difficult to attract either private or government funds. Government expenditures have generally been justified on the basis of downstream benefits, rather than improvement of range productivity per se.

Geologic erosion is a natural part of native ecosystems. It is the removal and deposition process that levels mountains and, in the past, has been responsible for the formation of the stable, gently sloping plains that are so vital to the world's agriculture. A big problem is the distinction between geologic and accelerated erosion.

Fig. 1. Soil mantle completely eroded except where protected by trees.

The problem of rangeland abuse and associated accelerated erosion is worldwide. Lowdermilk (1975), in his classic "Conquest of the Land Through 7000 years," which was first printed in 1953, attributed the downfall of several ancient civilizations to soil erosion and siltation problems resulting from mismanagement of the soil resources, primarily rangeland.

Historic abuse of western rangeland is fairly well documented (Box, 1979). Recent rangeland inventories indicated that much of it is in a deteriorated condition. Cutler (1980) stated that in the 48 contiguous states, 54% of all rangelands were in "poor or very poor condition with vegetation and soil conditions estimated to be at or less than 40% of their potential." Dregne (1978), in discussing desertification and its consequences on arid and semiarid rangelands, cited overgrazing as the main cause of land deterioration and loss of productivity. Overgrazing is especially devastating when rangelands are under drought stress. Most lands in the western half of the USA have undergone slight to very severe desertification, which is defined as the process of impoverishment of terrestrial ecosystems from man's activities. He stated that where land is very badly degraded the process of desertification is economically irreversible. This means that a considerable portion of the rangeland is covered by a vegetation complex that is more sparse and less protective than that which existed before the advent of white man.

It has generally been assumed that changes in vegetation cover due to overgrazing have been negative in terms of soil erosion protection. This is not always true. In the northern Great Plains, for example, the climax vegetation associated with range sites in excellent condition is often dominated by cool-season midgrasses, such as western wheatgrass (*Agropyron smithii* Rydb.) and needle-and-thread grass (*Stipa comata* Trin. and Rupr.) that produce considerable forage but provide only a limited amount of ground cover. As these sites are overgrazed, they become dominated by shortgrasses, such as blue grama [*Bouteloua gracilis* (Willd. ex H.K.B.) Lag.], and buffalograss [*Buchloe dactyloides* (Nutt.) Engelm.], threadleaf sedge (*Carex filifolia* Nutt.), and clubmoss (*Selaginella densa* Rydb.), that provide a large amount of ground cover and relatively little grazeable forage. It is difficult to establish a base for distinguishing between geologic and accelerated erosion where man-caused changes in vegetation have affected erosion rates. Factors other than grazing, such as prolonged drought, can result in significant cyclic changes in the vegetation cover.

If we want to recognize geologic erosion as a natural and important on-going process in native ecosystems, then we should base rangeland T values on the amount of geologic erosion from a specific range site in good-to-excellent condition under proper management. This is also consistent with sustained maximum production criteria. A rangeland in excellent condition will, at least theoretically, be at its maximum level of plant productivity. Because much of the rangeland has been abused and is in less than excellent condition, management for T values somewhat less than the current geologic rate may be needed to restore these lands to a production level consistent with their environment. Many rangeland soils are shallow and immature and their productivity potential, even in excellent condition, is greatly limited. Establishing T values less than the geologic erosion rate would accelerate the development of these soil profiles to the point that they would not be a limiting factor in plant growth. Some control of geologic erosion is feasible with treatments such as contour furrowing and water spreading, and with establishment of improved plant species.

It is difficult to establish T values for rangelands when the errors in soil-loss measurements may be in excess of the T values that should be established. Equally important is the fact that we only vaguely understand the effect of soil loss on sustained production, especially for rangelands. Until direct cause and effect relationships and reversibility and irreversibility are better understood, T values for rangeland may be a concept with only an idealistic application. Soil is the basic resource for food production. Its conservation and protection should be the controlling factors in all management schemes. Presently, however, it appears that rangeland management will have to be based on vegetation parameters, assuming that improving range condition reduces soil loss. While not always true, such an assumption provides a workable system for managing native rangelands to maximize sustained production.

LITERATURE CITED

Blackburn, W. H. 1980. Universal soil equations and rangelands. p. 588–595. *In* Symp. on Watershed Management 1980, Vol. I, 21–23 July 1980, Boise, Idaho. Am. Soc. Civil Engin., New York, N.Y.

Box, T. W. 1979. The American rangelands: Their condition and policy implications for management. p. 16–22. *In* Rangeland policies for the future, Proc. Symp., 28–31 Jan. 1979, Tucson, Ariz. Forest Service GTR-WO-17.

Branson, F. A., G. F. Gifford, and J. R. Owen. 1972. Rangeland hydrology. Range Science, Series No. 1, October 1972. Soc. for Range Manage., Denver, Colorado.

Cutler, M. R. 1980. Shaping a new commitment to the future of rangelands. Rangelands 2: 5–8.

Dregne, H. E. 1978. Desertification: Man's abuse of the land. J. Soil Water Conserv. 1(33): 11–14.

Kothman, M. M., Chairman, Range Term Glossary Committee. 1974. A glossary of terms used in range management. 2nd ed., Soc. Range Manage.

Langbein, W. B., and S. A. Schumm. 1958. Yield of sediment in relation to mean annual precipitation. Trans. Am. Geophys. Union 39:1076–1084.

Lowdermilk, W. C. 1975 (revised). Conquest of the land through 7,000 years. USDA, SCS, Agriculture Information Bull. no. 99.

Renard, K. G. 1980. Estimating erosion and sediment yield from rangeland. *In* Symp. on Watershed Management 1980, Vol. I, 21–23 July 1980, Boise, Idaho. Am. Soc. Civil Eng., New York, N.Y.

Ritchie, J. C., J. A. Spraberry, and J. R. McHenry. 1974. Estimating soil erosion from the redistribution of fallout ^{137}Cs. Soil Sci. Soc. Am. Proc. 38:137–139.

Skidmore, E. L., and N. P. Woodruff. 1968. Wind erosion forces in the United States and their use in predicting soil loss. USDA Handb. No. 346.

U.S. Department of Agriculture, Resources Conservation Act Coordinating Committee. 1980a. Soil and Water Resources Conservation Act: Program report and environmental impact statement 1980, Review Draft. USDA.

———. 1980b. Soil and Water Resources Conservation Act: 1980 Appraisal, Review Draft, Part I. USDA.

U.S. Department of Agriculture, Soil Conservation Service. 1975. Erosion, sediment and related salt problems and treatment opportunities. Special Projects Division, Golden, Colorado, December, 1975.

———. 1978. 1977 SCS national resource inventories.

Wadleigh, C. H. 1968. Wastes in relation to agriculture and forestry. USDA Misc. Publ. 1065.

Wight, J. R. 1976. Land surface modifications and their effects on range and forest watersheds. *In* H. F. Heady, D. H. Falkenborg, and J. P. Riley (ed.) Watershed management on range and forest lands. Proc. of the 5th Workshop of the United States/Australia Rangelands Panel, 15–22 June 1975, Boise, Idaho. Utah Water Research Laboratory, College of Engineering, Utah State Univ., Logan, Utah.

Wischmeier, W. H., and D. D. Smith. 1978. Predicting rainfall erosion losses: A guide to conservation planning. USDA Handb. no. 537.

Chapter 7
Technology Masks the Effects of Soil Erosion on Wheat Yields — A Case Study in Whiteman County, Washington[1]

H. A. KRAUSS AND R. R. ALLMARAS[2]

ABSTRACT

Sustained productivity of an eroding soil cannot be determined unless yield increases from technology advances are separated from soil productivity changes due to erosion. This concern is especially paramount in the Palouse landscape of Whitman County where serious erosion occurs under an intensive dryland wheat production, that has also had significant technology advances. The separation of technology and soil productivity involved the use of long-term wheat yields, measured wheat response to remaining epipedon, historical soil erosion rates, and landscape-distributed soil erosion rates. Current wheat yield in Whitman County increased approximately 36.1 kg/ha (0.54 bu/acre) per year as an average for the whole landscape; meanwhile annual soil erosion losses average 21.1 metric tons/ha (9.4 tons/acre) on a cropland base of 421,200 ha (1,040,000 acres). The soil productivity decrease from an average epipedon loss of 13.8 cm (5.3 in) in a 90-year period was 725 kg wheat/ha (10.8 bu/acre). An average erosion rate, however, does not reveal the true impact on productivity. Isolation of the soil productivity change component by land capability subclass showed that the net increase in yield on IIe and IIIe land (67% of the cultivated cropland) has masked a significant decline in productivity of subclasses IVe and VIe land (18% of the cultivated cropland) in the 90-year period of intensive cultivation. The average soil erosion rate in Whitman County over the 1940 to 1978 period has been nearly twice the tolerance value of 11.2 metric tons/ha (5 tons/acre) per year. Average annual soil erosion at the soil-loss tolerance (T value) level is expected to expose the subsoil of IVe land (about 12% of the cultivated cropland) in about 128 years.

---

[1] Contribution from the Soil Conservation Service, USDA, Spokane, Washington; and ARS, USDA, Pendleton, Oregon.
[2] State conservation agronomist, Spokane, WA 99201, and research soil scientist, Pendleton, OR 97801.

Copyright © 1982 ASA, SSSA, 677 South Segoe Road, Madison, WI 53711. *Determinants of Soil Loss Tolerance.*

INTRODUCTION

Much has been written (Horner et al., 1944 and 1960; Kaiser et al., 1954; and USDA and Washington State, 1979) about upland soil erosion in Whitman County and the Palouse River Basin (Fig. 1), but its impact on sustained productivity has not been determined. The estimated average annual soil erosion in Whitman County since 1930 is 21.1 metric tons/ha (9.4 tons/acre) from 421,200 ha (1,040,000 acres) of cultivated cropland. This average soil loss is almost double the 11.2 metric tons/ha (5 tons/acre) tolerance value established for deep, renewable soils. Historical accounts by the Soil Conservation Service suggest that soil erosion has occurred almost every year since intensive cultivation began about 1890.

Non-uniform seasonal distribution of precipitation in the Pacific Northwest requires a deep soil profile to store moisture for dryland agriculture. Annual precipitation in Whitman County (Fig. 1) ranges from about 33 to 58 cm (13 to 23 in), with about 70% received from November through June. Because transpiration needs are high and rainfall

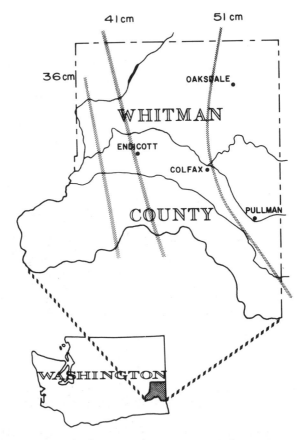

Fig. 1. Geographic location and precipitation in Whitman County.

Table 1. Yields of wheat in Whitman County, Washington for each decade since 1936.

Decade	Wheat yield (kg/ha)
1936–1945	2,029
1946–1955	2,305
1956–1965	2,984
1966–1975	3,474
40-year increase	1,445
Annual increase	36.1

is sparse from July through October, a deep soil profile is required for storage of the winter precipitation. About 90% of the soils supporting cultivated cropland are Mollisols, which have a mollic epipedon (Soil Conservation Service, 1975). These mollic epipedons are rich in plant nutrients, and they also have the high infiltration rates needed to recharge the soil profile during winter.

Some soils have had sufficient erosion to produce an epipedon < 18 cm (7 in) thick; these are ochric epipedons. We will use the term epipedon to include either a mollic or an ochric epipedon.

Three cropping sequences dominate cultivated cropland in Whitman County: winter wheat (*T. aestivum* L. ssp. *vulgare* and *T. aestivum* L. ssp. *compactum*) and summer fallow, wheat and peas (*Pisum sativum* L.), or wheat and lentils (*Lens culinaris* Medik). Winter and spring barley (*Hordeum vulgare* L. or *H. distichon* L.) are often grown in these sequences.

Summerfallow has created an especially serious erosion hazard in the higher rainfall zone because of the antecedent moisture stored in the soil profile. The many tillage operations used to control weeds also modify the surface soil structure and residue. The combination of winter precipitation and intermittently frozen ground often causes severe erosion when rain falls on snow covered or bare frozen ground (McCool et al., 1976). Summer fallowed land is especially sensitive to these winter conditions.

The need for erosion control might become more clearly recognized if the effects of erosion on yield could be quantified. To quantify these effects, one must separate the reduction in soil productivity from increases in wheat production produced by technology advances. This case study shows how technological effects were separated from soil erosion effects on wheat production in Whitman County. For this analysis we used information on long-term wheat yield trends, wheat response to epipedon thickness, and historical soil erosion rates. We also distributed soil erosion among interactive soil erosion capability subclasses in the Palouse landscape.

Wheat Yield Trends

Wheat yield trends for Whitman County (Table 1) were obtained from annual tabulations provided by the State Statistical Office[3] for the

---
[3] Kitterman, J. M. Printed annually. "All Wheat" Summaries. USDA-ESCS, State Statistical Office, Seattle, WA 98174.

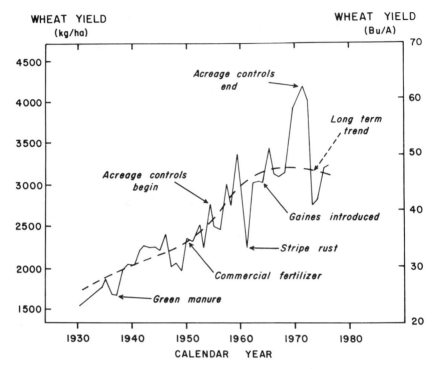

Fig. 2. Winter wheat production and technology inputs in Whitman County from 1930 to 1979.

1939-to-1975 period. Data for the 1936-to-1938 period were obtained from local ASCS[4] records. A plot of yield vs. time (Fig. 2) indicates the various technological inputs and their impact on production. Points on the statistically fitted trend line has a standard error of 380 kg/ha (5.6 bu/acre). The slope of this trend line has a standard error of approximately 5.4 kg/ha per year (0.08 bu/acre per year). Wheat yields increased 1,445 kg/ha (21.5 bu/acre) over the 40-year period from 1936 to 1975. The annual yield was 36.1 kg/ha (0.54 bu/acre).

Production increases from green manure in the late 1930s were evident until sweetclover weevil (*Sitona cylindricollis* Fahraeus) seriously reduced sweetclover (*Melilotus alba* Medik) production and discouraged the use of green manure (Fig. 2). Use of commercial fertilizers in the 1950s produced a general yield increase, which was sustained by the introduction of 'Gaines' and other semidwarf wheats in 1963. Farmers summerfallowed during the era of acreage controls because it was the best way to maximize wheat yields. Two notable yield declines occurred when stripe rust (*Puccinia striiformis* West.) attacked susceptible wheat cultivars in 1958, and when the end of acreage controls in 1973 dramatically reduced summerfallowing and encouraged cultivation of marginal land.

---

[4] Anonymous. Local unpublished records of USDA/ASCS, Colfax, WA.

The projected annual wheat yield increase in Whitman County was only 9% greater than the 33.6 kg/ha (0.5 bu/acre) per year, which Jensen (1978) ascribed to technological improvement over the 1935-to-1975 period in New York State.

Wheat yields shown in Table 1 are an average based on field conditions obtained under a mixture of managements and soils. These yields include a net effect of technological improvements and declining soil productivity (Fig. 2). Further analysis is necessary to document and separate these opposing trends.

Wheat Yield Responses to Epipedon Thickness

Wetter[5] determined a relationship between wheat yield and thickness of the epipedon. His measurement of the depth at which the brown organic matter coloration was no longer evident, or when structure or texture changed markedly, are both approximate measurements of the depth of the epipedon comprising the A and B horizons of Mollisols, Inceptisols, or Entisols in the Palouse. Over a period of 5 years (1970 to 1975), he made 90 paired measurements of winter wheat yield and associated epipedon thickness. There were also yield measurements in peas and lentils but they will not be discussed here. Wheat followed summerfallow, peas, or lentils. Measurements were made in central and northern Whitman County where the annual precipitation ranges from 41 to 58 cm (16 to 23 in). Approximately 90% of the Whitman county cropland is in this precipitation range. All measurements were made in farm fields, most of which received about 90 kg N/ha (80 lb/acre) preceding wheat planting. Preceding peas or lentils about 20 kg P/ha (18 lb/acre) and 15 kg S/ha (13 lb/acre) were applied. Wheat yield responses can be sensitive to losses of both nutritional and soil water depth as epipedon thickness is reduced.

Winter wheat yields were a linear function of measured epipedon thickness ranging from 0 to 61 cm (24 in). Statistical regression analyses produced coefficients of determination, indicating that at least 43% of the grain yield variations could be accounted for by thickness of the epipedon. The average standard error of estimated wheat yield was 1,200 kg/ha (18 bu/acre. As epipedon thickness ranged from 0 to 61 cm (24 in), wheat yield (averaged for observations after summerfallow, peas, or lentils) ranged from 2,352 to 5,645 kg/ha (35 to 84 bu/acre). The wheat yield decrease per cm of epipedon thickness reduction was 54 kg/ha (2.04 bu/acre per in). This estimate had a standard error of 10 kg/ha per cm (0.4 bu/acre per in).

The survey by Wetter[5] also showed similar declines in crop residue production as epipedon thickness was reduced. Wheat straw production after summerfallow declined from 11.8 metric tons/ha (5.3 tons/acre) with an epipedon 61 cm (24 in) thick to 3.5 metric tons/acre (1.6 tons/ha) when the epipedon was absent. Corresponding wheat straw productions

---

[5] Wetter, Fred. 1977. The influence of topsoil depth on yields. Tech. Note 10. Soil Conservation Service, Spokane, WA 99201.

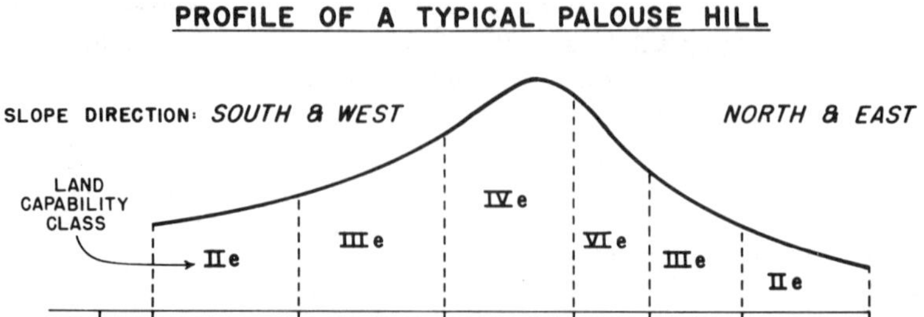

PROFILE OF A TYPICAL PALOUSE HILL

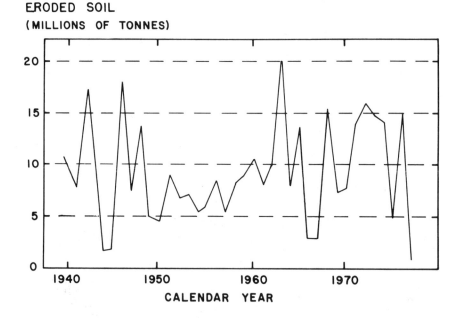

Fig. 3. Profile of a typical Palouse hill showing differences in percent slope, depth of epipedon, and soil organic matter for the constituent land capability subclasses. (Revised after Pawson et al., 1961).

Fig. 4. Sheet and rill erosion in Whitman County (421,200 ha) from 1940 to 1977.

after peas or lentils were 8.2 and 3.9 metric tons/ha (3.6 and 1.7 tons/acre). As residue production declines it is increasingly difficult to control soil erosion.

Fig. 3 shows the cross section of a typical Palouse hill as determined in 1961 (Pawson et al., 1961). The effects of epipedon loss on

Table 2. Land capability subclasses and their estimated contribution to soil erosion in Whitman County.

Slope (%)	Capability subclass†	Hectares	Percent of cultivated cropland	Percent of sheet and rill erosion§
3-7	IIe	16,848	4	2
7-25	IIIe	265,356	63	46
3-25‡; 25-40¶	IVe	50,544	12	25
25-40‡; 40-55#	VIe	25,272	6	27
Total all e		358,020	85	100

† Capability subclass based on slope and erosion hazard from Whitman County Soil Survey Report, Issued—April 1980.
‡ Eroded phase.
§ From Palouse Cooperative River Basin Study, USDA and Washington State, 1979.
¶ Walla Walla silt loam, 15 to 30% slopes are included in IVe.
# Walla Walla silt loam, 30 to 40% slopes are included in VIe.

productivity may be qualitatively related to the organic matter content which ranges from 1.8% in soil on the crest of the hill to 4.0% in the depositional area at the toe of the north and east slopes. Epipedon thickness typically ranges from less than 11 cm (4.3 in) in the hilltop position to more than 60 cm (23.6 in) at the toe of the south and west slope.

Soil Erosion Rates and Distribution Among Land Capability Subclass

From 1940 to 1978 the average annual soil erosion rate in Whitman County on a cropland base of 421,200 ha (1,040,000 acres) was estimated to be 21.1 metric tons/ha (9.4 tons/acre). The annual erosion ranged from 19.9 million metric tons (21.9 million tons) in 1963 to 0.49 million metric tons (0.54 million tons) in 1977 (Fig. 4). This information was provided by a study[6] in which rill erosion was measured on 10 to 20 key fields each year by the Alutin method or estimated in as many as 1,500 fields. The Alutin method measures rill cross-sections in a 418-cm (13.7-ft) transect perpendicular to rill direction. The Whitman County average annual soil erosion rate is less than half the rate of 44.8 metric tons/ha (20.0 tons/acre) measured in the Wild Horse Creek Watershed in northeastern Oregon during the period 1890 to 1940 (Thomas et al., 1943). Soils in the Wild Horse Creek Watershed are similar to those in Whitman County.

The distribution of rill and sheet erosion among land capability subclasses for the Palouse River Basin was assumed to apply to Whitman stitutes 72% of that in the Palouse River Basin. Of the 15 river basin evaluation areas used to gather on-site data, 13 were located in Whitman County. Distribution of soil erosion among land capability subclasses was obtained as the product of area of each soil map unit and a soil erosion rate both obtained in the same study. Subclass "e" cultivated cropland

---

[6] Kaiser, Verle. 1940-1978. Annual erosion survey of Whitman County. USDA-SCS, Spokane, WA 99201.

Table 3. Soil erosion losses and wheat productivity changes (from epipedon loss and technology input) in cultivated cropland as related to soil capability subclass in Whitman County.

Characteristic	Land capability subclass				Whitman County average
	IIe	IIIe	IVe	VIe	
Avg. erosion rate (metric tons/ha per year)	11.2	15.7	44.8	96.3	21.1
Years to lose 1 cm	12.6	9.0	3.1	1.6	6.7
Years to lose remaining‡ epipedon from a typical Palouse hill.	768	345	32	80	--
Soil loss in 90 years§ (cm)	7.1	10	29	56	13.4
Productivity change in§ 90 years (kg/ha)	−384	−541	−1,569	−3,040	−725
	(58)¶	(100)	(290)	(600)	(134)
Net productivity after adjusting for 1,446 kg/ha increase due to technology (kg/ha)	1,061	904	−124	−1,595	−720
	(58)	(100)	(290)	(600)	(134)

† Based on soil bulk density of 1.4 gm cm⁻³.
‡ See Fig. 3 for average remaining depth of epipedon in each land capability subclass for a typical Palouse hill.
§ The 90-year period is suggested based on the following quotation in USDA and Washington State (1979): "Summerfallow became a well established practice on most Palouse farms by the early 1890's. Washington and Idaho Experiment Stations began to recognize erosion as a problem." Productivity change is based on an average yield loss of 54.1 kg/ha per cm (2.04 bu/acre per inch of topsoil loss).
¶ A value in parenthesis is the standard error of the value just above in the same column.

area estimates for Whitman County in Table 2 were determined from the Whitman County Soil Survey Report.[7]

A set of average soil erosion rates by land capability subclass was obtained from on-site data derived from each of 13 soil associations located in Whitman County. Each evaluation area in the soil association comprised at least 600 ha (1,500 acres). All factors in the Universal Soil Loss Equation (USLE) (Wischmeier and Smith, 1978), including slope length and slope steepness, were estimated from observed field conditions. This equation predicts an average erosion rate for a given slope. The erosion rate on lower segments of uniform slopes or on steeper portions of complex slopes will exceed this average. The Whitman County slope transects were segmented and erosion rates corresponding to the various land capability subclasses were determined. The segmenting procedure thus reveals the actual distribution of erosion of each portion of the slope. The method of Foster and Wischmeier (1974) was used for segmenting based on the USLE adaptations of McCool et al. (1976) for the Pacific Northwest. Because this method accounts for upslope influences on soil erosion, it accounts for the interaction of land capability classes in their respective landscape positions.

---

[7] Soil Conservation Service. 1980. Soil survey, Whitman County, Washington. USDA-SCS, Spokane, WA 99201.

Technology Inputs Mask Soil Productivity Loss

Annual soil erosion rates by land capability subclass in Whitman County (Table 3) were computed using the area in each "e" subclass, the percent of erosion in Table 2, and the annual erosion measured in the Kaiser[6] study.

County average annual soil erosion rates varied from a low of 11.2 metric tons/ha (5 tons/acre) on subclass IIe land to a high of 96.3 metric tons/ha (43 tons/acre) on subclass VIe land (Table 3). Based on these estimated soil loss rates, 2.54 cm (1 in) of soil can be lost in 4 years on subclass VIe land and in 32 years on subclass IIe land. On class VIe land 30.5 cm (1 ft) can be lost in 48 years. Erosion rates for each capability subclass were projected in Table 3 to show how much epipedon has been removed from each subclass in the past 90 years.

Estimates of productivity loss (Table 3) due to soil erosion were based on a relationship discussed above between productivity and epipedon thickness. A severe decline in productivity due to epipedon loss is shown for subclass VIe land; some decline in productivity is shown even for subclass IIe land. After adjusting for the increased productivity from technology there was still a 1,595 kg/ha (24 bu/acre) projected productivity decline for subclass VIe land. Meanwhile, only a small loss of 124 kg/ha (1.4 bu/acre) net productivity was estimated for subclass IVe land. The gains of productivity attributed to subclasses IIe and IIIe land were enough to project a net productivity gain of 720 kg/ha (10.7 bu/acre) for Whitman County.

Because of technology gains since 1940, average wheat yields for Whitman County have not shown an overall decline. Even though an epipedon may have developed during the past 90 years, it has not offset a yield decline. Thus, average yields and average erosion rates may seriously conceal actual yield declines, especially on steep IVe and VIe subclasses. Table 2 shows that IVe and VIe subclasses constitute only 18% of the cultivated cropland but produce 52% of the soil erosion.

Significant abandonment or land-use change on subclass VIe land has not occurred because: a) farm production and cost/return data by capability subclass is lacking, b) cropland base or Normal Conserving Acres was feared lost if seriously eroded areas were seeded to grass, and c) these steep eroding areas are recognized as low-producing land suitable for set-aside (idle land in the farm program).

Epipedon losses from subclasses IVe and VIe land not only reduce crop productivity but also create management difficulties. Earlier we showed that thinner epipedons produced less crop residue with which to control soil erosion and improve soil tilth. Because poorer structured subsoil or parent material is exposed when the epipedon is eroded, infiltration is expected to decrease, runoff to accelerate, and soil water storage to decrease. Other expected problems are decreased rooting depth, more soil crusting of seedbeds, and nonuniform machinery response.

In Table 4, the current wheat yield distribution by land capability subclass on a Palouse silt loam is compared with wheat yields adjusted for

Table 4. Comparison of current wheat yield distribution on a Palouse silt loam with the distribution predicted from epipedon loss in Table 3.

Land capability subclass	Current wheat yields† on Palouse silt loam (kg/ha)	Avg.‡ Whitman County (kg/ha)	Land class adjustment§ (kg/ha)	Net predicted yield (kg/ha)
IIe	5,040	3,474	1,061	4,539 ± 58¶
IIIe	4,032	3,474	904	4,378 ± 100
IVe	3,360	3,474	−124	3,350 ± 290
VIe	1,747	3,474	−1,595	1,879 ± 600

† Source: USDA/SCS. 1975. Yields and erosion rates by soils in Whitman County. Soil Conservation Service, Spokane, Washington 99201. Wheat yield on Palouse silt loam subclass VIe land obtained from the Wetter report cited earlier.
‡ Average during the 1966 to 1975 decade (from Table 1).
§ Adjustment from Table 3.
¶ Standard error of predicted yield.

epipedon loss (Tables 1 and 3). Estimates based on Table 3 exceeded those current in Whitman County for capability subclass IIIe by only 356 kg/ha (5.3 bu/acre). The current yield (column 2, Table 4) exceeded the estimates for IIe, IVe, and VIe by 430, 20, and 87 kg/ha (6.4, 0.3, and 1.3 bu/acre), respectively. Predicted yields for each capability subclass based on the data presented in the paper closely parallel current actual yields.

Government farm programs and acreage controls strongly influenced erosion control in at least 18 of the last 40 years. Subclass VIe land could have been protected with permanent grass or alfalfa, or at least a rotation that periodically included these protective, soil-building crops. Farmers who practiced this conservation, however, lost "crop history". Farm programs also encouraged summerfallowing, which is extremely detrimental to subclass VIe land in the intermediate and high precipitation zones of the Palouse River Basin (USDA and Washington State, 1979).

Even though green manuring (sweetclover) reduced soil erosion in the 1930s, it could not be continued because of clover weevil damage, incompatibility with herbicides, commercial fertilization, and competition with the wheat nurse crop. Current alternatives for erosion control are elimination of summerfallow, reduced production of peas and lentils, use of minimum tillage and no-till, continuous spring grain or rotation of spring grains with winter grains, and permanent or rotational grass/legume cover on hilltops.

Economic Effects of Soil Loss

The 1976 crop year was used for a comparison of actual and potential production (Table 5) in Whitman County; 1977 was a drought year and 1978 was abnormally wet. If soil productivity had not declined, farm sales of wheat in 1976, alone, could have been 22% greater. The percentage of land in each "e" subclass for Whitman County was used to distribute 85% of the 218,700 ha (540,000 acres) of wheat harvested in Whitman County in 1976. In 1976 there were 723,741 metric tons (797,775 tons) of wheat produced in Whitman County, or an average of

Table 5. Current wheat yields in 1976 in Whitman County compared to potential wheat yields had soil productivity not been reduced by 90 years of soil erosion.

Characteristic	Land capability subclass			
	IIe	IIIe	IVe	VIe
Area (ha)	8,748	137,781	26,244	13,122
Productivity loss (kg/ha)	384	541	1,569	3,040
Wheat loss (1,000 metric tons)	3.35	73.15	40.39	40.47

Total 1976 deficit:	157,354 metric tons or 719 kg/ha (10.7 bu/acre)
Actual 1976 total:	723,741 metric tons or 3,306 kg/ha (49.2 bu/acre)
Potential 1976 without soil productivity decline:	881,095 metric tons or 4,025 k/ha (59.9 bu/acre)
Direct loss at $104.71/deficit metric ton:	$16,478,110

3,306 kg/ha (49.2 bu/acre). Had soil productivity not degraded, it is estimated that production would have been 881,095 metric tons or an average of 4,025 kg/ha (59.9 bu/acre). At a price of $104.71/metric ton ($2.85/bu), $16,478,110 were lost in 1976, alone. When other related losses are considered, such as adverse effects of sediment on crop yields in bottomlands, road repair, water quality, and recreation, the losses as a result of soil erosion are immense.

Soil Loss Tolerance

An interesting extension of our analysis is to project epipedon and productivity loss using an annual erosion rate equal to soil loss tolerance. For deep, renewable soils the tolerance value is 11.2 metric tons/ha per year (5 tons/acre per year). Shallow nonrenewable soils have tolerance values less than 11.2 metric tons. The tolerance value is the standard used by the Soil Conservation Service when evaluating alternative conservation systems.

This annual erosion rate of 11.2 metric tons/ha for 50 years would remove 4.0 cm (1.5 in) of epipedon with an estimated yield decline of 208.3 kg/ha (3.1 bu/acre). In 100 years at 11.2 metric tons/ha, about 7.9 cm (3.1 in) of epipedon would be removed with an estimated yield decline of 423.4 kg/ha (6.3 bu/acre). An average erosion rate, however, does not reveal the true impact on productivity. After about 128 years of erosion at an average rate of 11.2 metric tons/ha per year, there would be no epipedon left on the more erosive subclass IVe land, which constitutes about 12% of the cultivated cropland in Whitman County.

CONCLUSIONS

According to Jensen (1978), "It seems unlikely that any future combination of genetics, technology, or unknown factors will be able to generate a sustained rise in productivity from those (present) high levels. . productivity will continue to grow, but at a slower rate." Our estimates of productivity losses take on much more significance if Jensen's appraisal is correct.

The declining yield of existing eroding cropland must also be evaluated in terms of the decreasing agricultural land base as more and more prime cropland is converted to other uses.

It will be much more difficult in the future for technology to mask declining yields. This is particularly true if erosion rates continue at their present level or accelerate in the years ahead. Soil loss tolerance specifications must be at realistic levels if we are to realize the benefits of increased productivity due to improved technology. Yield declines due to erosion may offset or even outweigh increases resulting from technological advances.

It is necessary to provide more research information on the soil management needed to restore productivity of the soil already eroded enough to expose the subsoil.

Research and case evaluations are needed throughout the United States if we are to fully comprehend the long-term effects of soil erosion and develop a policy to deal with it.

LITERATURE CITED

Foster, G. R., and W. H. Wischmeier. 1974. Evaluating irregular slopes for soil loss prediction. Am. Soc. Agric. Eng. Trans. 17:305-309.

Horner, G. M., A. G. McCall, and F. G. Bell. 1944. Investigations in erosion control and reclamation of eroded land at the Palouse Conservation Experiment Station, Pullman, Washington, 1931-1942. USDA. Tech. Bull. 860. 83 p.

————, M. M. Oveson, G. O. Baker, and W. W. Pawson. 1960. Effects of cropping practices on yield, soil organic matter, and erosion in the Pacific Northwest wheat region. Pacific Northwest Agric. Exp. Stn. Res. Ser. Bull. 1.

Jensen, N. F. 1978. Limits to growth in world food production. Science 201:317-320.

Kaiser, V. G., W. W. Pawson, M. H. H. Groeneveld, and O. L. Brough, Jr. 1954. Soil losses on wheat farms in the Palouse wheat-pea area, 1952-1953. Washington Agric. Exp. Stn. Circ. 255. 11 p.

McCool, D. M., R. I. Papendick, and F. L. Brooks. 1976. The Universal Soil Loss Equation as adapted to the Pacific Northwest. p. 135-147. In Proc. 3rd Fed. Inter-Agency Sedimentation Conf. Water Resources Council, Washington, DC.

Pawson, W. W., O. L. Brough, Jr., J. P. Swanson, and G. M. Horner. 1961. Economics of cropping systems and soil conservation in the Palouse. Agric. Exp. Stn. of Idaho, Oregon and Washington, and ARS, USDA Bull. 2. 82 p.

Soil Conservation Service. 1975. Soil taxonomy: a basic system of soil classification for making and interpreting soil surveys. USDA Agric. Handb. 436. 754 p.

Thomas, H. L., R. E. Stephenson, C. F. Freese, R. W. Chapin, and W. W. Huggins. 1943. The economic effect of soil erosion on wheat yields in eastern Oregon. Oregon Agric. Exp. Stn. Circ. 157. 32 p.

USDA and Washington State. 1979. Palouse cooperative river basin study. Soil Conservation Service, Spokane, WA 99201. 182 p.

Wischmeier, W. H., and D. D. Smith. 1978. Predicting rainfall erosion losses; a guide to conservation planning. USDA Agric. Handb. 537. 58 p.

Chapter 8

Soil Loss Tolerance[1]

E. L. SKIDMORE[2]

ABSTRACT

A function is developed for defining soil loss tolerance (T value) that provides for permanent preservation of the soil resource, prohibits erosion that contributes excessively to pollution, and is a function of the present soil depth. The relationship is expressed by $T(x,y,t) = (T_1 + T_2)/2 - (T_2 - T_1)/2 \cos[\pi(z-z_1)/(z_2-z_1)]$, where $T(x,y,t)$ is tolerable soil loss rate at point (x,y), and T_1 and T_2 are lower and upper limits of allowable soil loss rate, T_1 corresponds to soil renewal rate, z_1 and z_2 are minimum allowable and optimum soil depths, and z is the present soil depth. Tolerable soil loss function between the points (T_1,z_1) and (T_2,z_2) is sinusoidal and dependent upon soil depth and $(T_2 - T_1)/2$ is the amplitude. The period is represented by the cosine argument and goes from 0 to 180 degrees for values of z between the limits of z_1 and z_2. Examples of application are given.

INTRODUCTION

The soil on which we depend for existence is a limited resource. The potential gross cropped area accessible to relatively high-yielding cultivation with present technology is about 4.2 billion ha (Revelle, 1976), of which one-third to one-half of the most productive part is already under cultivation. Arable land is limited not only in area but also in depth, quantitatively and qualitatively. Many processes can reduce the soil's current and/or potential capacity to produce desired crops. Processes that degrade soils include desertification, wind erosion, water erosion and sedi-

---

[1] Contribution from ARS-USDA, in cooperation with the Kansas Agric. Exp. Stn. Dep. of Agronomy Contribution 79-296-J.
[2] Soil scientist, ARS, USDA, Manhattan, Kansas 66506.

Copyright © 1982 ASA, SSSA, 677 South Segoe Road, Madison, WI 53711. *Determinants of Soil Loss Tolerance.*

mentation, flooding, waterlogging, organic matter oxidation, physical deterioration, chemical pollution, salinization, alkalinization, and urbanization.

Concern that soil degradation processes be stopped or reversed has prompted comments and actions of various kinds. Lowdermilk (1951) suggested an 11th Commandment on stewardship of the land.

Food and Agriculture Organization of the United Nations (1974, 1977) recently sponsored two expert consultations for assessing soil degradation. At the first of these, the panel recommended that land be recognized as an essential and limited resource and that the adverse effect of different forms of soil degradation on future food suppliers of the world be considered. They suggested that the highest priority be given to soil conservation measures urgently needed to ensure the supply of an adequate diet for increasing populations.

At a recent board meeting[3], the Soil Science Society of America approved a resolution that land and water conservation programs be instituted that will conserve our vital soil resource, maintain its productivity, and foster a healthful environment.

A reasonable objective of soil stewardship is to maintain a soil resource that with judicious use of additional available resources of water, favorable climate, plants, and technology can produce sufficient food and fiber to meet the present and future needs of man on Earth. Obviously, accomplishing this objective depends not only on the soil resource itself but also on other resources that enhance soil productivity and the demand created by the world population. However, we will limit the following discussion to considerations of soil loss tolerance on cultivated cropland. Our objective is to develop a usable function for defining tolerable soil loss which includes the concepts developed by Stamey and Smith (1964).

They suggested that the definition of erosion tolerance must (1) provide for the permanent preservation or improvement of the soil as a resource, (2) be adaptable to the erosion and renewal rates of any soil characteristic, (3) be a function of position since at any two points on the Earth's surface the erosion and renewal rates will not necessarily be identical, (4) be applicable regardless of the cause of erosion or renewal, and (5) allow the use or depletion of any soil property (e.g., depth) in excess of present or predictable future requirements.

Conceivably, erosion of some soils such as a deep loess soil could cause more serious environmental problems than impairment of a soil resource. This suggests the need for another element in the definition of soil loss tolerance: (6) prevent erosion from contributing excessively to pollution and other environmental problems.

METHODS

Stamey and Smith (1964) developed a mathematical expression for erosion tolerance at point (x,y) of some measurable soil property.

$$I(x,y) - \int_{t_0}^{\infty} [E(x,y,t) - R(x,y,t)] \, dt \geq M(x,y) \qquad [1]$$

---

[3] See minutes of SSSA Board of Directors' Meeting, 7 Dec. 1978, Chicago, Ill.

Table 1. The various combinations of limits for the curves of Fig. 1.

Curve	T_1	T_2	z_1	z_2
	——— mm/yr† ———		——— m ———	
a	0.2	2.0	0.5	0.8
b	0.2	2.0	0.5	1.2
c	0.2	2.0	0.5	1.6
d	0.2	2.0	0.5	2.0
e	0.2	1.6	0.5	2.0
f	0.2	1.2	0.5	2.0
g	0.2	0.8	0.5	2.0

† When a soil has a bulk density of 1.0 g/cm³, multiply millimeters per year by 10 to convert to metric tons per hectare per year.

where $I(x,y)$ is the position function, which gives the value of the measure of the soil property at the initial time (t_0; $M(x,y)$ represents the minimum allowable value at (x,y) of the measure of this property; and $E(x,y,t)$ and $R(x,y,t)$ represent the erosion rate and renewal rate, respectively, of the measurable soil property. Equation 1 defines the concept that net change tolerance $[E(x,y,t) - R(x,y,t)]$ integrated over time subtracted from the initial value of the measurable soil property must always exceed the minimum allowable value. However, very little progress has been made in the last 15 years to define the function of Eq. [1] so that it is of practical use.

Consider the following equation for defining tolerable degradation of some measurable soil property at point (x,y). For illustration and discussion, let us apply the equation to soil depth, although it could be used for other measurable properties, both extensive and intensive.

$$T(x,y,t) = (T_1 + T_2)/2 - (T_2 - T_1)/2 \cos[\pi(x - z_1)/(z_2 - z_1)] \quad [2]$$

Where T_1 is the lower limit of allowable rate of change of soil property at point (x,y) (here it represents soil loss per annum); $T(x,y,t)$ equals T_1 when soil depth is at minimum allowable value so that net change function of Eq. [1] equals zero. In other words, $\int_{t_0}^{t_1} [E(x,y,t) - R(x,y,t)]dt$ equals zero and $P(x,y)$ equals $M(x,y)$, where $P(x,y)$ is the present value of the soil property depth.

The upper limit of allowable soil loss rate at point (x,y) is T_2; $T(x,y,t)$ equals T_2 when soil depth is great enough so that a further increase in soil depth would not further enhance the productive capacity of that soil at point (x,y); z is the present value $P(x,y)$ of the soil property at point (x,y); z_1 is the minimum allowable value of the soil property at point (x,y), [$M(x,y)$ of Eq. 1]; and z_2 is the optimum or target value $O(x,y)$ of the soil property at point (x,y). At this point, increasing the value of z_2 would not further increase the productive capacity of that soil at point (x,y). $T_1 \leq T(x,y,t) \leq T_2$ as $M(x,y) \leq P(x,y) \leq O(x,y)$. The relationship between the limits of T_1 and T_2 for allowable soil loss and z_1 and z_2 for soil depth is sinusoidal. The amplitude is $(T_2 - T_1)/2$.

Period is represented by $[(z - z_1)/(z_2 - z_1)]\pi$ of the cosine argument and ranges from zero to π radians or 180 degrees for values of z between the limits of z_1 and z_2. The first term of Eq. [2] is simply an amplitude offset. $T(x,y,t)$ connects the extreme points (T_1,z_1) and (T_2,z_2) with a slope of

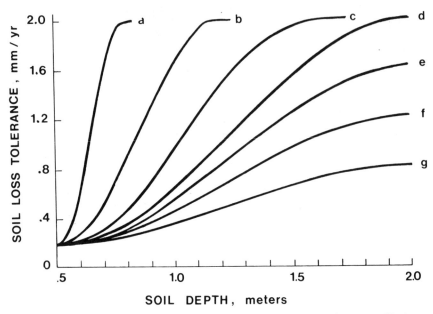

Fig. 1. Soil loss tolerance as a function of soil depth for various combinations of limits.

zero. That is, as the present value $P(x,y)$ of the soil property depth gets closer to $M(x,y)$, the change in $T(x,y,t)$ with change in $P(x,y)$ goes to zero and $T(x,y,t) = R(x,y,t)$.

Soil loss tolerance as a function of soil depth, as defined by Eq. [2] for various combinations of limits (Table 1), is illustrated by Fig. 1. Each curve shows the values of T for the full period or half cycle of the cosine function. The lower limits were held constant. The minimum allowable soil depth $M(x,y)$ was chosen at 0.5 m; renewal rate $R(x,y,t)$ was chosen at 0.2 mm/year.

APPLICATION EXAMPLE

Suppose we wished to determine appropriate soil loss tolerances for a soil at (x,y) that was 1.4 m deep. We judged that the production capacity of that soil would increase with depth up to 2.0 m and that a depth of 0.5 m would be the minimum allowable. Soil renewal rate is 0.2 mm/year. We determined that maximum soil loss should never exceed 2.0 mm/year. Then, using Eq. [2], we calculated tolerable soil loss as 1.38 mm/year, or about 14 metric tons/ha•year.

When different values for present value of soil depth are substituted into Eq. [2], curve d of Fig. 1 is generated. As the soil depth approaches either limit of soil loss tolerance, the slope of the curve approaches zero.

Now we must answer the questions: How fast does the soil depth change with time if soil depth changes according to soil loss tolerance (T value), and in what manner does T value change with time? This was done by solving Eq. [3] and substituting P(x,y) for z back in Eq. [2] for n = 2,000 iterations.

$$P_i(x,y) = I(x,y) + \sum_{i=0}^{i=n-1} [R_i(x,y,t) - T_i(x,y,t)] \qquad [3]$$

where the variables are as defined previously. The results are shown in Fig. 2. Soil depth decreases rather quickly with time from the initial value of 1.4 m, then levels off with time as the T value approaches the soil renewal rate (0.2 mm/year).

Now, for another example, suppose that a very deep soil of uniform depth had 0.5 m soil that could be removed without affecting its current or future productivity. Furthermore, we assumed that z_1 and z_2 are 1.0 and 1.5 m, respectively; and T_1 and T_2 are 0.1 and 2.0 mm/year, respectively. In this case, Eq. [2] and [3] yield the results shown in Fig. 3. Soil loss tolerance remains constant at the maximum value until soil depth equals 1.5 m, then decreases as soil depth decreases below 1.5 m.

DISCUSSION

In these applications, reasonable values for the upper and lower limits of T value and soil depth were assumed. The key to the successful use of Eq. [2] in describing soil loss tolerance lies in the rationale and procedure for determining limit values. Some may argue that we do not need the upper limit of T value because only the lower limit is important in permanently preserving the soil resource. However, knowing the upper limit is important for meeting criterion No. 6. Clean Air Amendments (1970) and Federal Water Pollution Control Act Amendments (1972) will not allow us to permit wind and water erosion to occur without consideration of the environment. Also, we should guard against soil loss where the damage costs to the environment are greater than the costs of preventing the loss.

Figures 2 and 3 indicate that soil depths change rather slowly with time for the conditions of these examples. In these cases, the renewal function was constant. We need more information on renewal rates for specific locations and conditions and how renewal rates can be accelerated. Renewal rates from weathering of the basalt underlying shallow loess soils in the Pacific Northwest is slow as compared with those of shale parent material of some soils in the Southeast. We should not permit soil loss to proceed to the extent that it lowers the producing capacity of the soil, either immediately or in the long-term, more than it would cost to prevent the loss.

The minimum allowable soil depth could be defined in terms of the present soil depth and/or, according to the judgment of local soil scientists, the depth of soil required to preserve high-level crop production. Results now being obtained on reclaiming drastically disturbed lands should give additional insights into depth and quality of soil needed for particular production levels.

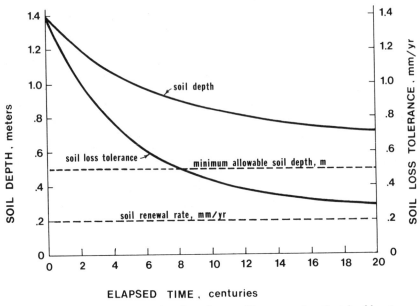

Fig. 2. Change in soil depth and T value as soil loss proceeds at the tolerable rate.

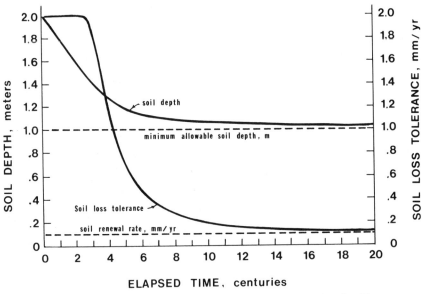

Fig. 3. Change in soil depth and T value as soil loss proceeds at the tolerable rate.

The optimum soil depth could also be defined in terms of the present soil depth or according to the judgment of soil scientists who know local conditions and limiting resources. In addition to meeting T values, we should increase the quantity and quality of the soil resource wherever the cost of doing so is lower than the value of the increased production capacity.

So far we have discussed the soil property depth as if it were measurable, which it is. However, we must establish some guiding criteria for that measurement. Do we measure A + B horizons, rooting depth of commonly grown crops, depth to impervious layer, or something else, and do we establish a weighting factor for quality of soil material giving more credit to the more desirable topsoil? These questions and those pertaining to defining limits can be answered best by consensus of concerned and knowledgeable scientists representing various groups like the Soil Conservation Service, Agricultural Research, and Cooperative State Research Service.

With appropriate limits, Eq. [2] would satisfy criteria for defining erosion tolerance mentioned earlier and be very useful for determining T value. Then, using the WEQ (Woodruff and Siddoway, 1965; Skidmore and Woodruff, 1968) and the USLE (Wischmeier and Smith, 1978), we can implement erosion control practices to maintain a soil resource that can produce food and fiber to meet our present and future needs.

LITERATURE CITED

Clean Air Amendments of 1970. 1970. Public Law 91-604. U.S. Code Congressional and Administration News. 1954-2001.

Federal Water Pollution Control Act Amendments of 1972. 1972 Public Law 92-500. U.S. Statutes at Large 86:816-903.

Food and Agriculture Organization of the United Nations. 1974. A world assessment of soil degradation—an international programme of soil conservation. Report of an expert consultation on soil degradation, Rome. 40 p.

————. 1977. Assessing soil degradation—report of an FAO/UNEP expert consultation, Rome, FAO Soils Bull. 34. 83 p.

Lowdermilk, W. C. 1951. Conquest of the land through seven thousand years. USDA, SCS, MP-32. 38 p.

Revelle, Roger. 1976. The resources available for agriculture. Sci. Am. 235:164-178.

Skidmore, E. L., and N. P. Woodruff. 1968. Wind erosion forces in the United States and their use in predicting soil loss. USDA, ARS, Agric. Handb. No. 346. 42 p.

Stamey, W. L., and R. M. Smith. 1964. A conservation definition of erosion tolerance. Soil Sci. 97:183-186.

Wischmeier, W. H., and D. D. Smith. 1978. Predicting rainfall erosion losses—a guide to conservation planning. USDA, Agric. Handb. No. 537. 58 p.

Woodruff, N. P., and F. H. Siddoway. 1965. A wind erosion equation. Soil Sci. Soc. Am. Proc. 29:602-608.

Chapter 9

Current Criteria for Determining Soil Loss Tolerance

D. E. McCORMACK, K. K. YOUNG, AND L. W. KIMBERLIN[1]

ABSTRACT

Current criteria for determining soil loss tolerance (T value) are imperfect. Much is still unknown about the rates at which a favorable root zone forms and about the effects of erosion on soil productivity. It is known, however, that erosion at the rate of the maximum tolerance of 11.2 metric tons/ha/year exceeds the estimated rate at which most parent rocks weather.

Water quality objectives are too variable to provide a basis for determining T value. Additional work on the sediment delivery ratio is needed to clarify the relationship between soil erosion and water quality. A sediment limit is proposed as an aid in planning conservation systems that can limit soil erosion to meet water quality objectives.

Policies on the amount of erosion that can be tolerated should not be made solely by the scientist. Farmers and other land users, soil and water conservation districts, special interest groups, and political groups must participate in developing such policies. A major responsibility of the scientist is to inform these policy makers of the likely consequences of various options available to them. Political expediency and short-sighted environmental or economic demands cannot be allowed to determine tolerable levels of soil erosion.

---

[1] Staff leader and soil scientist, Soil Technology Staff; and national agronomist, Retired, Soil Conservation Service, Washington, D.C.

Copyright © 1982 ASA, SSSA, 677 South Segoe Road, Madison, WI 53711. *Determinants of Soil Loss Tolerance.*

INTRODUCTION

If soil erosion were to occur for several centuries at the rate of currently accepted values of soil loss tolerance (T value), substantial losses in productivity would result on many of our soils. The rate of soil formation simply will not compensate for such a rate of soil erosion, and thus the total rooting zone would gradually become thinner and of lower quality.

Criteria for determining T value are now being re-evaluated by the SCS and cooperating agencies. The criteria will need to be changed if the central objective in conservation planning is to maintain the productivity of soils in perpetuity.

To date, the question of T value has been approached primarily as a technical issue. We have studied the maintenance of soil productivity, the prevention of gullies, and the control of off-site sediment damage. These three factors have been the main considerations in current criteria for determining T value; but SCS is studying the need to change them. We have also studied the results of failure to manage our soils to meet these needs. These studies have been of great value and further technical study is needed.

How much erosion can be tolerated? The issues surrounding this question are not limited to technical ones. Ethical issues are also at stake, and some of these are extremely difficult to state clearly enough for soil scientists, agronomists, or any others to resolve them. The difficulty stems mostly from the fact that we do not know exactly what demands will be placed on our soil resources in the future.

USDA is developing a detailed analysis of soil and water conservation needs and progress. Reports are due to Congress and the President in late 1981 and again in 1985, as required by the Soil and Water Resources Conservation Act (RCA) of 1977 (P.L. 95-192). Thus, the recent new attention to the question of soil loss tolerance is timely.

Review of Early Work

The fact that soil erosion robs the land of its productivity has been known for a long time. This fact has been stated dramatically by McDonald (1941) and Lowdermilk (1953) and was the core of eloquent arguments advanced in the late 1920s for the Federal Government to develop programs that would reduce erosion damage to our land (Bennett, 1939). Because of these arguments, the role of the Federal Government in erosion prevention was increased. Public awareness of the problem gradually increased between the time of organization of the SCS in 1935 and the first Earth Day (22 Apr. 1970). Since that time, public interest and concern among nonfarm groups and individuals have increased sharply.

The idea that the kind of management required for erosion control varied with kind of soil was well stated by Bennett (1939): "Effective prevention and control of soil erosion . . . requires the use and treatment of all the various kinds of land . . . in accordance with the individual needs and adaptabilities of each. . . ."

By the mid-1930s there were estimates that total sediment carried to the sea from our land was about 3.3 billion metric tons annually (Bennett, 1939). However, specifying a T value could be meaningful only after a method was developed to estimate the amount of soil being eroded.

CONCEPT OF TOLERANCE

Reference to specific amounts of soil erosion that might be tolerated on a specific soil occurred just after early attempts by Zingg (1940) to predict amounts of soil erosion. In considering the soil losses on Fayette silt loam, Hays and Clark (1941) concluded that erosion should be kept under 6.7 metric tons/ha/year. They reasoned that at this annual rate it would take about 50 years to remove 2.5 cm and that the top 20 cm would be lost in about 400 years. From a practical standpoint, they felt that farmers would regard such a loss as reasonably safe.

Hays and Clark (1941) noted that to produce 30 cm of soil from limestone requires the solution of about 30 m of rock, and thousands of years. They concluded that a loss even as small as 6.7 metric tons/ha/year would greatly exceed the annual production of new soil from limestone.

In the same year, Smith (1941) reviewed data from Bethany, Mo., and concluded that the acceptable rate of soil loss would vary with the type of soil treatment, if all other factors were constant. He cited unpublished data indicating that soil productivity was increasing slightly on a plot in a 3-year rotation of corn, wheat, and meadow with soil treatment and a soil loss of 9.0 metric tons/ha/year. On a plot with the same rotation but without soil treatment, productivity was decreased and erosion was about 11.2 metric tons/ha/year.

Smith (1941) observed that a loss of 9.0 metric tons/ha/year may be too high to maintain the productivity of an eroded Shelby soil in which the surface soil is diluted with subsoil. However, he also observed that the addition of organic matter might overcome the loss of productivity due to erosion, and he even speculated that the dilution of the surface soil with a small amount of subsoil could be beneficial.

In a thorough review of soil erosion research in the Midwest, Browning et al. (1947) identified, for 12 soils, the "maximum average annual permissible soil loss without decreasing productivity." They observed that a given soil erosion loss is more serious on soils with claypans or heavy infertile subsoils than on loessial soils with uniform texture throughout the profile. The soils of Iowa were placed into six groups of soil erodibility characteristics. The main factor used in determining soil loss tolerance was the loss of productivity per 2.5 cm of soil lost through erosion. The early soil loss tolerance values of Browning et al. (1947) and similar data from other workers are presented in Table 1.

SOIL LOSS EQUATIONS

In the late 1940s, soil conservationists from the former SCS regional office in Milwaukee, especially C. L. Parish, recognized the value of the

Table 1. Early and current soil loss tolerance values for selected soils.

Soil series or phase	Soil loss tolerance value	
	Early	Current
	ton per hectare/year	
Fayette silt loam (9)	6.7	11.2
Fayette silt loam (3)	9.0	11.2
Ida silt loam (3)	13.5	11.2
Marshall silt loam (3)	11.2	11.2
Sharpsburg silt loam (3)	11.2	11.2
Clarion silt loam (3)	9.0	11.2
Clinton silt loam (3)	9.0	11.2
Storden silt loam (3)	6.7	11.2
Shelby silt loam (3)	6.7	11.2
Shelby silt loam (20)	9.0	11.2
Weller silt loam (3)	6.7	6.7
Dodgeville (deep) silt loam (3)	4.5	9.0
Rockton (deep) (3)	4.5	9.0
Malvern (3)	4.5	
Dubuque silt loam (3)	4.5	6.7

early soil loss equation for farm planning and teamed with research workers in the region to develop a system for regional application (Smith, 1961). Under the direction of G. W. Musgrave, a nationwide workshop was held in 1946 in Cincinnati, and all soil loss data obtained by that date were reviewed (Musgrave, 1947). The SCS regional office in Milwaukee then developed instructions for field use of the soil loss equation for conservation farm planning. Soil loss tolerances were assigned to key soils in the Midwest by 1948 based on the judgment of the best informed soil scientists.

In Illinois, "Guide for the Management of Soils, Field Crops and Pastures" was published cooperatively by the Illinois Agric. Exp. Stn. and SCS in 1950. It included T values and indicated the cropping systems suitable for different soils, slopes, and erosion control practices. Extensive data on crop yields in Illinois (Odell, 1950) were used as a basis for judging the impact of erosion on soil productivity, a key factor in determining T values.

During the 1950s, additional data were collected to help document the deleterious effects of soil erosion. A number of meetings were conducted, mostly in the Midwest, to discuss the Universal Soil Loss Equation (USLE) and T value. The soil loss that might be considered tolerable on soils with thick, favorable subsoils was a subject of lively debate. In the early 1950s, a maximum of 15.7 metric tons/ha/year was advocated by L. J. Bartelli and B. Boatman, SCS State Soil Scientists in Illinois and Iowa, respectively. The Tama soil was chosen to represent the deeper soils on which erosion would have the least effect on yields. Data were presented to show that corn yield was reduced on Tama soils by 0.27 q/ha-cm of topsoil lost (Van Doren and Bartelli, 1956). These authors placed a 10.1 metric ton/ha/year tolerance on Tama soils. They observed that for Fayette soils the yield reduction was 0.37 q/ha-cm of topsoil lost and, by simple factoring, deduced that the soil lost tolerance for Fayette soils should be 7.9 metric ton/ha/year.

In Iowa in the early 1950s, the T value for soils such as Tama was set at 13.5 metric tons/ha/year (Thompson, 1952). For shallow soils on bedrock, the tolerance value was 1.1 metric tons/ha/year in Iowa and 3.4 metric tons/ha/year in Illinois (Van Doren and Bartelli, 1956).

All reference to maximum soil loss tolerance for Tama and similar deep soils after the late 1950s indicate that 11.2 metric tons/ha/year was reasonable, until the mid-1970s when SCS began to reevaluate the issues (Young, 1978).

Current Criteria for Soil Loss Tolerance

Soil loss tolerance is defined as the maximum rate of annual soil erosion that will permit a high level of crop productivity to be obtained economically and indefinitely. This definition has been used since the early 1940s when the application of soil erosion equations made it possible to predict the amounts of soil erosion. The definition strongly implies—but does not directly state—that there should be no loss in the long-term productive capability of the soil. Long-term productivity depends on maintaining the thickness and quality of the A horizon and a sufficient favorable rooting depth.

HISTORY OF SCS GUIDELINES

Guidelines for T value were formulated in the early 1960s after about 15 years of discussion by soil scientists, agronomists, and others. In 1961 and 1962, SCS held six regional soil loss prediction workshops attended by soil scientists, agronomists, geologists, and soil conservationists. At these workshops, guidelines were developed for establishing T value, the 11.2-metric tons/ha/year maximum tolerance was set, and T values were set for the major soils in the region.

At the SCS regional workshops, the idea was advanced that erosion rates of more than 11.2 metric tons/ha/year tended to lead to the formation of gullies while lower rates did not (Klingebiel, 1961). This generalization was no doubt recognized to be as marginally valuable then as it is now, because no single erosion rate can be related to gully formation for all kinds of soil, climate, crops, or tillage methods.

Current SCS guidelines for T value were distributed by SCS in 1973 in an attachment to Advisory Notice Soils-6 (see Appendix). This advisory notice requested the SCS State staffs to update the soil erodibility (K) and T factors according to these guidelines.

Other guides have been developed by the four SCS technical service centers. For example, the following rules are used in States served by the Northeast Technical Service Center (Garland, 1962).

—Soils with loamy sand or sand textures to depths of 120 cm have a tolerance of 11.2 metric tons/ha/year.

—Soils with moderately coarse and finer textures and no impeding layer (fragipan, claypan, or bedrock) and soils with no coarse sand and gravel within 90 cm have a tolerance of 9.0 metric tons/ha/year.

—Soils with an impeding layer (fragipan, claypan, or bedrock) or coarse sand and gravel between 30 and 90 cm of the surface have a tolerance of 6.7 metric tons/ha/year. Severely eroded phases of these soils have a tolerance of 4.5 metric tons/ha/year.

—Soils with an impeding layer (fragipan, claypan, or bedrock) or coarse sand and gravel within 30 cm of the surface have a tolerance of 4.5 metric tons/ha/year.

The procedure used in assigning T value has relied strictly on multiple judgments of informed scientists (Edwards, 1962). Stamey and Smith (1964) proposed an equation to derive the tolerances, but it was not adopted because of the difficulty in quantifying the variables.

Criteria currently used to determine T value relate to maintaining soil productivity. Before 1977, prevention of off-site damage by sediment was commonly cited as a basis for the 11.2 metric tons/ha/year maximum T value. Since 1977, exclusion of this factor from the determination of T value has been under study by SCS because, for many watersheds, soil erosion limits that might be required to meet water quality standards differ widely from those necessary for maintaining soil productivity.

MAINTAINING SOIL PRODUCTIVITY

The maintenance of soil productivity depends on management practices and soils and site characteristics, all of which are considered in setting T values. The major characteristics are soil rooting depth, topsoil thickness, available water capacity, plant nutrient storage, surface runoff, soil tilth, and soil organic matter content.

Loss of the A Horizon. Loss of the A horizon was studied in numerous research projects in the 1930s and 1940s that investigated the general idea that loss of the A horizon might reduce yields much more on some soils than on others. After early reports on the impact of erosion on yields by Smith (1941) and Browning et al. (1947), many additional studies were made. In some studies, the entire A horizon or portions of it were removed by grading and yield levels were measured on the resulting soil (Follett et al., 1974). An extensive review of corn, wheat, and grain sorghum yields as influenced by topsoil thickness (Lyles, 1975) indicates greater yield reductions resulting from erosion than had been reported earlier. For example, for the Tama soil, Lyles (1975) reported a corn yield reduction of 0.79 q/ha-cm of topsoil lost compared to 0.27 reported in 1956 (Van Doren and Bartelli, 1956). Lyles (1975) reports corn yields from six studies and notes a range in yield reduction from 0.74 to 1.69 q/ha-cm of topsoil lost, compared with a range of about 0.25 to 0.74 q/ha-cm reported on similar soils by Browning et al. (1947). It is likely that a substantial part of this difference is due to the much higher base yield levels in 1975 than in the earlier studies.

Yield impacts of a similar magnitude were reported in western Tennessee (Overton and Bell, 1974). On several soils, corn yields over a 16-year period averaged 66 q/ha on uneroded soils and 45 q/ha on their severely eroded counterparts. If we assume that the A horizon is 25 cm thinner in the severely eroded soil, yields have been reduced by 0.84 q/ha-cm of topsoil lost. Recent economic analyses show that farm profits fall off rapidly because of the loss in productivity that results from erosion (Seitz et al., 1979).

The reduction in available water capacity is no doubt the main reason for the lower productivity of most eroded soils. But it has also been shown that runoff is greater on eroded soils than on uneroded soils; therefore, less water infiltrates into the soil for use by the crop and yields are reduced (Wischmeier, 1966). This yield reduction is greatest in areas of frequent moisture stress. Loss of organic matter not only reduces available water capacity but also reduces nutrient storage.

For soils to be productive, it is important that the tilth of the plow layer be maintained. Loss of seedings and poor stands of cultivated crops result from erosion. Both of these considerations have been cited as important in establishing T values. They are the main reasons why, in Mollisols with heavy B horizons such as the Swygert soils, productivity loss per cm of soil lost in higher in the lower part of the A horizon than in the upper part (Odell, 1948).

Another factor in determining T value that relates to maintenance of soil productivity is the value of the plant nutrients that are lost per ton of soil eroded. In 1962, these losses were estimated at $2 per ton of soil (Edwards, 1962). Recent work indicates that 0.9 metric ton of good topsoil contains 1.4 kg of N, 1.1 kg of P, and 20 kg of K (Willis and Evans, 1977). At 1979 prices for fertilizers, these nutrients have a value of $5.81. Most farmers would certainly consider an annual loss of $72/ha excessive (erosion equal to T value of 11.2 metric ton/ha/year).

From sketchy data it was estimated that, in permeable, medium textured material in well-managed cropland, an A horizon can form at the rate of 2.5 cm in 30 years (Bennett, 1939). This formation rate, if it is assumed that 6.27 ha-cm of soil weighs 165 metric tons, is equivalent to the formation of 11.2 metric tons/ha of new topsoil each year. This estimated rate of A horizon formation is the single most important reason that maximum T value has been established at that level.

Loss of Favorable Rooting Depth. The weathering of parent rock into a favorable root zone is a distinctly different phenomenon from the formation of the A horizon. In most soils it proceeds much more slowly. As a result, limiting annual soil erosion to the T value of 11.2 metric ton/ha might maintain the A horizon thickness for many centuries but the total root zone would become thinner.

Data on the rate of development of a favorable root zone from weathering of parent material are not yet conclusive, although numerous reviews have been made of reports on the rate of rock weathering (Smith and Yates, 1968; Smith and Stamey, 1965; Young, 1978) and dust deposition (Smith et al., 1970). A renewal rate of 1.1 metric tons/ha/year is thought to be a useful average for unconsolidated materials. For most consolidated (rock) materials, rates are much lower.

Keying the rate of permissible loss of favorable rooting depth to the rate of soil renewal will assure that soil thickness is maintained. In practice, however, on most cropland it would be extremely difficult—if not impossible—to limit erosion to 1.1 metric ton/ha/year.

In Fig. 1 and 2, two sets of T values and a soil renewal rate of 1.1 metric ton/ha/year are used to show the decrease in root zone thickness

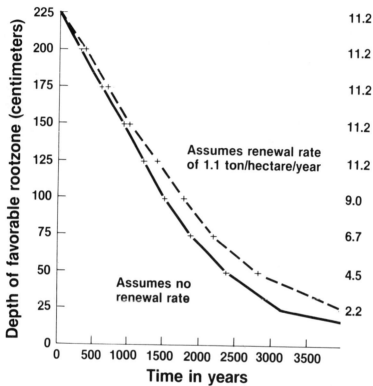

Fig. 1. Time required to decrease rootzone depth where soil loss tolerance is 11.2 metric ton/hectare/year initially and declines as indicated.

over several thousand years. In both figures it is assumed that the rate of erosion equals the T value and that value is reduced as the thickness of the root zone decreases. In Fig. 1, initial soil loss tolerance is 11.2 metric tons/ha/year and is reduced when depth of favorable root zone is 40 inches. In Fig. 2, the initial soil loss tolerance is 22.5 metric tons/ha/year initially and is reduced stepwise with the loss of each 24 cm of soil.

According to Fig. 1, a cropland soil with 150 cm favorable for plant growth would decrease to 100 cm in 660 years; a soil 225 cm thick would decrease to 100 cm in 1,800 years.

In Fig. 2, a soil with a 150 cm root zone with decrease to 100 cm in about 400 years. A soil with a 225 cm root zone would decrease to 100 cm in about 1,050 years.

There is little question that the 100 cm soil would be less productive for most crops than the 150 or 225 cm soil. Data are inadequate, however, to show conclusively how much soil productivity would decline over several centuries.

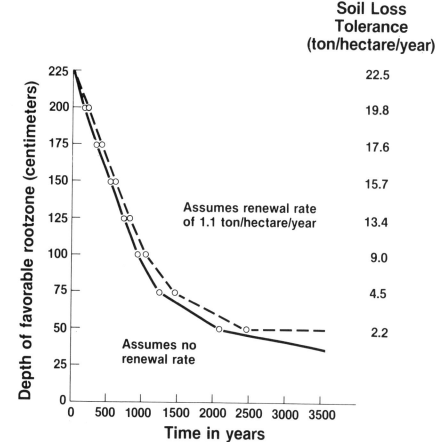

Fig. 2. Time required to decrease rootzone depth where soil loss tolerance is 22.5 metric ton/hectare/year initially and declines as indicated.

OFF-SITE DAMAGES DUE TO SEDIMENT

As the concept of T value was being adapted and refined in the early 1960s, it is clear that preventing off-site damage was one of the major bases for the criteria used. For example, Edwards (1962), after stating that maintaining productivity was an important objective, added "and without serious sedimentation." The main emphasis was given to siltation of waterways and terrace channels. As early as 1959, however, damage to lakes and reservoirs was cited among those factors used to determine the T value (Springer, 1959).

In 1976 and 1977, the prevention of off-site damage as a factor in determining T value was debated extensively by SCS and cooperating agencies. It was concluded that the control of soil erosion to prevent off-site damage depended on the water quality objectives in the specific watershed. To protect a trout fishery, a 2.2 metric ton/ha/year erosion

Table 2. Percentage of cropland experiencing stated erosion rates by major crops—1977 (Unpublished data, SCS 1977 National Resource Inventory).

Crop	Erosion rates, ton/hectare/year		
	<11.2	11.2–22.5	>22.5
Corn	67	17	16
Soybeans	56	26	18
Cotton	66	22	12
Sorghum	61	23	16
Wheat	87	9	4
Peanuts	53	32	15
Tobacco	47	26	27

rate might be excessive, while on another watershed with the same soils, erosion rates of 22.5 metric tons/ha/year would present no water quality problems. Thus there is no maximum tolerance value that could be assigned to individual kinds of soil to achieve water quality objectives.

A further difficulty is that the relationship between soil erosion rates and water quality is not well known. We just do not know very well where the eroded soil goes. A better understanding of the sediment delivery ratio is badly needed. In the meantime, SCS has concluded that preventing off-site sediment damage should not be a criterion for determining T value (Young, 1978). When sediment delivery ratios are better known, it will be possible to establish specific limits in soil erosion amounts that must be met on given soils in a watershed so that stated water quality objectives can be met (Wischmeier and Smith, 1978).

Future Application of Soil Loss Tolerance

Monoculture, the use of large farm machinery, increased population, conversion of cropland to urban uses, and high prices for crops are increasing the demands on our soils. In several parts of the nation, no approach to the production of cultivated crops is now known that can keep erosion within the T value and that is economically feasible to farmers. Many farmers will not apply conservation practices if they lose money because of doing so.

CURRENT EROSION LEVELS

The SCS National Resource Inventory of 1977 (USDA, 1978) shows that, on 27% of our cropland, erosion is occurring at a rate greater than the T value. On 23%, or nearly 40 million ha, of our 167 million ha of cropland, erosion exceeds 11.2 metric tons/ha/year. On 17 million ha of cropland, erosion exceeds 22.5 metric tons/ha/year. The inventory also indicates that T value is 11.2 metric tons/ha/year on two-thirds of our cropland. The extent of high soil erosion rates (Table 2) shows clearly that soil erosion is still a serious menace to our nation.

Because soil erosion not only lowers soil productivity but also increases the amount of energy required per unit of production, it is an energy tariff on future generations who may be less able to pay than we are today. Recent demands for energy have prompted studies on the removal of organic residues from the land for use in the generation of energy. Intensified demands on the land are likely to result. Soil scientists have the responsibility of identifying constraints and options calling attention to soil erosion and other hazards, and arguing for wise use of soils.

SOIL CONSERVATION OBJECTIVES

Soil conservationists have generally assumed that a basic guideline for conservation planning should be that the erosion rates should not exceed the T value. According to this line of reasoning, conservation objectives ideally must include maintenance of the long-term productivity of the soil.

In an ideal world, we would know the rate of formation of a favorable root zone for each soil. We would use tillage methods and other agricultural practices that would allow us to produce needed food and fiber while assuring that soil erosion does not exceed the rate of favorable root zone formation. In this ideal world, we would have full control of all aspects of agricultural production. We would know the carrying capacity of the land and the consequences of alternatives in its use and management.

Unfortunately, soil erosion on most cultivated sloping soils today exceeds the rate of formation of a favorable root zone. Given this fact, how should we state erosion control objectives so that they are reasonable both today and in the future? The question must not be decided on technical grounds alone, because they are not fully understood and are not likely to provide a clear-cut answer in the foreseeable future. Economical, ethical, and moral considerations must be included. A new land ethic and a new reverence for the land are needed (Brown, 1979).

Changing the Criteria for Soil Loss Tolerance. It is possible that T values could be set on the basis of technical factors alone, whereas soil conservation objectives could be set by political or social institutions to reflect ethical, economic, or social concerns.

The setting of T value, thus limited to a purely technical activity, would be free from implications about land use restrictions that in the past may have prevented a fully objective evaluation objectives. With the liberty thus provided, we could assign T values that do not result in the decline in productivity indicated by the curves in Fig. 1 and 2. If a favorable root zone is found to form at the rate of 1.1 metric tons/ha/year and if a reduction in thickness would decrease productivity, then we could assign a T value equal to the rate of formation of a favorable root zone.

Conservation objectives could then be based on a societal consensus on the need for production from land susceptible to erosion. Where the objectives would result in erosion at rates greater than the T value, the scientist must explain the long-term impacts on productivity. But perhaps it should be the political and social leaders who determine the objectives in soil conservation. To do so, they should be informed about the T value but not strictly bound by it. And they must have the technical insight to properly defend their action.

Sediment Limit. The sediment limit is proposed as the maximum permissible annual sediment yield that may be permitted from a given tract of land without causing water pollution in excess of defined water quality limits. After the total annual sediment yield limit for the watershed has been established, maximum erosion rates could be determined for the given tract of land. For example, in a 404 ha watershed, assume that a maximum of 900 tons of sediment per year can be permitted within water quality standards. Also assume the watershed has 161 ha of pastureland and woodland that, through the USLE, are found to be contributing 90 tons of sediment per year; one large gully is contributing 90 tons of sediment per year; and road ditches and other critical areas are contributing 180 tons per year.

The balance of the total annual sediment limit for the watershed, 540 tons would be permissible on the 243 ha of cropland in the watershed. The sediment limit for this cropland is, therefore, 2.24 metric tons/ha per year. The cropland conservation system would be designed to limit erosion to a rate determined by the sediment delivery ratio and the proportion of the 243 ha that is sloping land contributing sediment.

Needs for Information

Because there are a number of unknowns relative to soil loss tolerance—the greatest of which is a lack of knowledge about the rate of soil renewal—studies of soil erosion and its prevention need renewed emphasis. The scientific re-evaluation of assumptions and objectives should be given higher priority than it has had in the last decade or two. In addition, the following points should be debated both within and outside the scientific community:

—The rate of population growth affects the time horizon that would be appropriate in considering T value and in setting conservation objectives. We have not adequately justified a maximum of 11.2 metric tons/ha/year for T value on our thickest, most productive soils or the eventual destruction, in 2,000 or 3,000 years (Fig. 1 and 2), of their productivity. Whereas early work focused on the impacts of the loss of the A horizon, future work must concentrate on the effect of loss of the whole soil.

—The rationale for soil loss tolerance should be stated in terms that make it clear that conservation systems on specific kinds of soil will achieve valid objectives. It would be fortunate if those objectives were centered on maintenance of soil productivity in perpetuity.

—Objectives in conservation planning might state that damages from infrequent, severe storms, e.g., 100 year storms, during which as much as 100 to 250 metric tons/ha of erosion may occur, should be tolerated.

—Criteria could be developed for rating the quality of the root zone as a medium for plant growth. These criteria might be used in an arithmetic approach to deriving T value. These ratings might have many other applications, e.g., deriving yield estimates for crops not currently grown.

—The effect of soil erosion on water quality needs to be studied quantitatively so that soil erosion limits can be set that will achieve stated water quality goals. Because of the influence of particle size and the tendency of soil particles to stay in suspension, those limits would have to vary by kinds of soil. The limits would also vary by water quality objectives. Most of the chemical pollutants transported by runoff from farmland are adsorbed on clay particles; this fact is justification for tying soil loss limits for water quality to kinds of soil.

—The assumption that there is an upper biological limit to the food or fiber yield of a given crop needs to be studied further. If no such limit exists, is it really necessary to worry about maintaining productivity as we think of it today?

—Proper incentives must be found to convince landowners that controlling soil erosion is distinctly to their advantage. Soil conservation practices may reduce returns to the farmer (Seitz et al., 1979) in the short run, at least. Until proper incentives exist, our scientific endeavors to define T value may amount to little more than another casualty of the tragedy of the commons (Hardin, 1968), wherein the pursuit of personal gain makes it impossible to achieve a societal goal.

Stewardship and an Uncertain Future

More than 25 years ago safe minimum standards in conservation policy were discussed in social and economic terms by Ciriacy-Wantrup (1952). He noted that the irreversible depletion of resources could cause heavy social losses; the longer that resource depletion continues, the more difficult it is to halt and reverse. Soil scientists must learn more about the reversibility of the impacts of soil erosion. In particular, they must study and determine the rate at which the productivity of eroded soils can be restored and the kind of management required to restore it.

Just how irreversible are the impacts of erosion? On a given soil, projections based on current data indicate that average corn yields may be reduced by 0.74 q/ha/year for each 2.5 cm of soil loss. With an average annual erosion rate of 22.5 metric tons/ha, 2.5 cm of soil will be lost every 15 years. Over a 100-year period on a soil that has already lost the top 2.5 cm of soil, a productivity loss of 74 q/ha would result solely from the loss of that first 2.5 cm. Loss of the next 2.5-cm increment would further decrease production by 65.9 q/ha, or 0.78 q/ha/year for 85 years. And so on. Calculated over the 100-year period, during which the loss of 2.5 cm increments of soil would occur, the total loss in corn production would be 272.8 q/ha. Of course, such calculations are highly tentative, and they do not allow for possible technological breakthroughs that might offset productivity losses caused by erosion.

The problem of uncertainty and its implications relative to conservation policy are also discussed by Ciriacy-Wantrup (1952). There are a number of factors related to soils and erosion for which the future outlook is uncertain. Surely, future demands on our soils will exceed today's level and it seems likely that a point will be reached within less than 500 years when food production will control world population. To what extent will the total population of the earth depend on the adequacy of soil conservation from now on?

We are also uncertain about the extent to which an individual farmer should be responsible for preventing erosion. It is certain that future generations will benefit from erosion control and that they cannot pay for this benefit (Seitz and Hewitt, 1979). On the other hand, if we fail to control erosion in our time, future generations will not be able to avoid paying for the failure. We believe it is clear that, to a large extent, the significance of our work is to maximize the value of their heritage. Yet to the farmer it may be clear that it is possible, at little or no risk, to subject his soil to soil erosion levels that are excessive even from the standpoint of just two generations.

To cope with such uncertainty, we see no alternative but to adopt a very generally stated ethic of stewardship. We must strive to protect our soil and its productive capability and "leave it as an inheritance to our sons forever" (Ezra 9:12). In 1854, Chief Seattle admonished the white man who wanted to buy the land of his people. He asked them to "teach your children so that they will respect the land . . . that the earth is our mother . . . whatever befalls the earth befalls the sons of the earth" (Seattle, 1976). In the words of Lowdermilk (1953), "a starving farmer will eat his seed grain; you will do it and I will do it, even though we know it will be fatal to next year's crop. Now is the time, while we still have much good land capable of restoration to full or greater productivity, to carry through a full program of soil and water conservation."

No doubt we will learn a great deal more in the future about the rates of formation of favorable root zone in various kinds of parent materials, about methods of tillage that will hold soil erosion to very low levels, and about restoring the productivity of severely eroded soils. This prospect, however, should not lead us to relax our efforts to minimize erosion, because the possibility of irreversible depletion will remain.

SUMMARY

The current T values were based largely on considerations of the rate of formation of A horizons, with adjustments based on the thickness or other aspects of the quality of the entire soil depth accessible to plant roots. It is now clear that for most soils, T values are several times greater than the rates of soil information.

With only a few exceptions, we do not know the rates of soil formation nor the long-term impacts of soil erosion on crop yields. Consequently, there is an argument for basing T values on social and political objectives. However, the scientist is obliged to inform the politician of the consequences of alternative actions. The scientist is also obliged to study the rates of soil formation and the impacts of erosion to the extent that ultimately technical considerations can form an accurate basis for social goals to preserve the productivity of our soils.

APPENDIX

Guide for Developing the Soil Loss Tolerance Factor (T) in the Universal Soil Loss Equation

Soil loss tolerance (T), sometimes called permissible soil loss, is the maximum rate of soil erosion that will permit a high level of crop productivity to be sustained economically and indefinitely.

Soil loss tolerance values of 1 through 5 are used. The numbers represent the permissible tons of soil loss per acre per year where food, feed and fiber plants are to be grown. The T values are not applicable to construction sites or to other non-farm uses of the erosion equation.

A single T value is normally assigned to each soil series. A second T value may be assigned to certain kinds of soil where erosion has significantly reduced the thickness of the effective root zone; thus reducing the potential of the soil to produce plants over an extended period of time. For example, eroded phases of soil series that are shallow to moderately deep to a soil layer that restricts roots are commonly given a T value one class lower than the uneroded phase of the same soil. The following criteria are used by soil scientists and other specialists for assigning T values to soil series:

1. An adequate rooting depth must be maintained in the soil for plant growth. For soils that are shallow over hard rock or other restrictive layers, it is important to retain the remaining soil; therefore, not much soil loss is tolerated. The T value should be less on soils shallow to impervious layers than for soils with good soil depth or for soils with favorable underlying soil materials that can be renewed by management practices.

Guide for assigning soil loss tolerance values to soils having different rooting depths.

Rooting depth	Soil loss tolerance values (Annual soil loss—tons/acres)	
inches	Renewable soil†	Non-renewable soil‡
0–10	1	1
10–20	2	1
20–40	3	2
40–60	4	3
60+	5	5

† Soils with favorable substrata that can be renewed by tillage, fertilizer, organic matter, and other management practices.

‡ Soils with unfavorable substrata such as rock or soft rock that cannot be renewed by economical means.

2. Soils that have significant yield reductions when the surface layer is removed by erosion are given lower soil loss tolerance values than those where erosion effects yield very little.

A maximum of 5 tons of soil loss per acre per year has been selected for use with the USLE. This maximum value has been used for the following reasons:

1. Soil losses in excess of 5 tons per acre per year affect the maintenance, cost and effectiveness of water-control structures such as open ditches, ponds, and other structures affected by sediment.
2. Excessive sheet erosion is accompanied by gully formation in many places causing added problems to tillage operations and to sedimentation of ditches, streams and waterways.
3. Loss of plant nutrients. The average value of nitrogen and phosphorus in a ton of soil is about $2 to $3. Plant nutrient losses of more than $10 per acre per year is considered to be excessive.
4. Numerous practices are known that can be used successfully to keep soil losses below 5 tons per acre per year.

LITERATURE CITED

Bennett, H. H. 1939. Soil conservation. McGraw-Hill, Inc., New York.

Brown, L. R. 1979. Population, cropland, and food prices. Nat'l Forum, The Phi Kappa Phi J. LXIX(2):11-16.

Browning, G. M., C. L. Parish, and J. Glass. 1947. A method for determining the use and limitations of rotations and conservation practices in the control of erosion in Iowa. Agron. J. 39(4):65-73.

Ciriacy-Wantrup, C. V. 1952. Economics and policies. Univ. of California Press, Berkeley.

Edwards, M. J. 1962. Soil erodibility factor values and soil loss tolerance. In Soil loss prediction for South Carolina. Mimeo. SCS, USDA, Columbia, South Carolina 29201.

Follett, R. F., D. L. Grunes, G. A. Reichman, and C. W. Carlson. 1974. Recovery of corn yield on Gardena subsoil as related to leaf and soil analysis. Soil Sci. Soc. Am. Proc. 38(2):327-331.

Garland, L. E. 1962. Soil factor values and soil loss tolerance for local soils. In Soil loss prediction for the northeastern states. Mimeo. SCS, Broomall, PA 19008.

Hardin, G. 1968. The tragedy of the commons. Science 162:1243-1248.

Hays, O. E., and N. Clark. 1941. Cropping systems that help control erosion. State Soil Conserv. Committee Bull. 452, in cooperation with the Soil Conservation Service, USDA, and the Agric. Exp. Stn., Univ. of Wisconsin, Madison.

Klingebiel, A. A. 1961. Soil factor and soil loss tolerance. In Soil loss prediction, North and South Dakota, Nebraska, and Kansas. SCS, Lincoln, NE 68508.

Lowdermilk, W. C. 1953. Conquest of the land through seven thousand years. Agric. Inf. Bull. No. 99, SCS, USDA, Washington, DC 20250. 30 p.

Lyles, L. 1975. Possible effects of wind erosion on soil productivity. J. Soil Water Conserv. 30:279-283.

Musgrave, G. W. 1947. The quantitative evaluation of factors in water erosion, a first approximation. J. of Soil Water Conserv. 2(3):133-138.

McDonald, A. 1941. Early American Soil Conservationists. Misc. Pub. No. 449, SCS, USDA, Washington, DC 20250. 62 p.

Odell, R. T. 1948. Relationship between corn yield and depth of surface soil in Swygert silt loam in 1946 and 1947. AG 1373b, Dep. of Agronomy, Agric. Exp. Stn., Univ. of Illinois, Urbana.

―――. 1950. Measurement of the productivity of soils under various environmental conditions. Agron. J. 42:282-292.

Overton, J. R., and F. F. Bell. 1974. Productivity of soil on the west Tennessee experiment station for corn. p. 35-38. In Tenn. Farm and Home Sci. Prog. Rep. 90. Univ. of Tennessee, Knoxville. April-June 1974.

Seattle, Chief. 1976. If we sell you our land. Fellowship 42:3-7, 11.

Seitz, W. D., and R. Hewitt. 1979. Equity Analysis in Public Policy Formation. Project Rep. Contract No. 68-03-2597 EPA. Athens, GA 30605.

Smith, D. D. 1931. Interpretation of soil conservation data for field use. Agric. Eng. 22:173–175.

————. 1961. History of the soil loss estimation and the new equation. *In* Soil loss prediction, North and South Dakota, Nebraska, and Kansas. SCS, Lincoln, NE 68508.

Smith, R. M., P. C. Twiss, R. K. Krauss, and M. J. Brown. 1970. Dust deposition in relation to silty season and climatic variables. Soil Sci. Soc. Am. Proc. 34(1):112–117.

————, and W. L. Stamey. 1965. Determining the range of tolerable erosion. Soil Sci. 100(6):414–424.

————, and R. Yates. 1968. Renewal, erosion, and net change functions is soil conservation science. 9th Int. Congr. of Soil Sci. Trans. Vol. 5, paper No. 77.

Soil Conservation Service. 1970. K and T factors of soil series mapped in the northeast region. SCS, USDA, Broomall, PA 19008.

Springer, D. K. 1959. Soil loss tolerance—T. *In* Soil loss estimation in Tennessee. Mimeo. SCS, Nashville, TN 37203.

Stamey, W. L., and R. M. Smith. 1964. A conservation definition of erosion tolerance. Soil Sci. 97(3):183–186.

Thompson, L. M. 1952. Soils and soil fertility. 1st ed. McGraw-Hill, Inc., N.Y.

USDA 1978. 1977 National Resource Inventories. SCS, P.O. Box 2890, Washington, DC 20013.

Van Doren, C. A., and L. J. Bartelli. 1956. A method of forecasting soil loss. Agric. Eng. 37(5):335–341.

Willis, W. O., and C. E. Evans. 1977. Our soil is valuable. J. Soil Water Conserv. 32:258–259.

Wischmeier, W. H. 1966. Relation of field plot runoff to management and physical factors. Soil Sci. Soc. Am. Proc. 30(2):272–277.

————, and D. D. Smith. 1978. Predicting rainfall erosion losses—a guide to conservation planning. Agric. Handb. no. 537, USDA, Washington, DC 20250.

Young, K. K. 1980. Impact of erosion on soils of the United States. *In* M. DeBoodt and D. Gabriels (ed.) Assessment of erosion. John Wiley and Sons, New York. N.Y.

Zingg, A. W. 1940. Degree and length of land slope as it affects soil loss in runoff. Agric. Eng. 21(2):59–64.

Chapter 10

Erosion Tolerance for Cropland: Application of the Soil Survey Data Base[1]

R. B. GROSSMAN AND C. R. BERDANIER[2]

ABSTRACT

A proposal is given for the systematic assignment of soil loss tolerances (T values) based on properties of soil profiles. The T value is conceived as the product of a T value assigned to deep, uniform soils and a T value adjustment factor. The former presumably may change with societal needs. The proposed T-value adjustment factor is calculated from soil properties and would remain constant unless the criteria change. Potential plant rooting depth is pivotal to the T-value adjustment factor and involves a search to the 2-m depth for (1) taxonomic features indicative of root growth limitation; (2) horizons with high strength and weak structure; and (3) horizons high in extractable Al, low in Ca, or both. An additional adjustment is made based on soil property changes within the potential plant rooting depth that are indicative of the effect erosion may have on soil productivity.

INTRODUCTION

The following capsule definition of soil loss tolerance (T value) from Soil Conservation Service (SCS) working documents describes the concept we address: "Soil loss tolerance is the maximum rate of soil erosion that can occur without long-term reduction in the productive capacity of the soil." The amount of soil loss allowable can be approached systematically by separating the current T value into two quantities, the product of which is the T value. One is the T value (T-max) assigned to deep soils

---
[1] Contribution from National Soil Survey Laboratory, SCS, Lincoln, NE. Presented before Div. 5, 6, 7 of Soil Sci. Soc. Am., Fort Collins, Colorado, 8 Aug. 1979.
[2] Soil scientists, USDA, SCS, Lincoln, NE.

Copyright © 1982 ASA, SSSA, 677 South Segoe Road, Madison, WI 53711. *Determinants of Soil Loss Tolerance.*

having uniform properties with depth. The other quantity, referred to as the T-value Adjustment Factor (TAF), is a number from 0 to 1 that is determined by depth to a plant root limitation and certain changes with depth above that plant root limitation. In arithmetic terms then, T value = $T_{max} \times TAF$. We assume that T_{max} may change with need for food production and that determination of the values would involve a wide spectrum of interests. In contrast, TAF would be based strictly on soil properties and would be determined by technical people. If the proposal were accepted, TAF would replace the T value as the quantity of record attached to specific soils.

To explore the calculation of TAF, we first consider generally the soil survey data base, with emphasis on establishing the soils on which TAF would be determined and on obtaining plant root depth information. Second, we present a procedure to determine the potential rooting depth. And third, we propose a relationship for calculation of TAF that involves potential rooting depth and a Depth Change Factor (DCF). The latter is determined by changes from the soil surface to the potential rooting depth in selected soil characteristics.

The approach to be presented differs markedly from current practice in the National Cooperative Soil Survey and in the Soil Conservation Service. The proposals are offered for testing only and do not represent a policy change by the SCS.

THE DATA BASE AND ITS USE

Soil survey information is extensive with the most important source being the 1,000 or more people who map soils. Most cropland subject to erosion has been mapped. Modern soil surveys have been completed on 75% of the United States where 30% or more of the cropland has a major erosion limitation (Stewart et al., 1975, Fig. 7; the 1979 soil survey status map, Soil Survey Staff, 1979). Each of the 13,000 current soil series descriptions contains information on landscape position, morphology, and composition. These descriptions also have supplemental numerical estimates of bulk density, available water capacity, particle-size distribution, and permeability. For many soil series, detailed laboratory data for specific pedons are published in Soil Survey Investigations Reports and in publications of state laboratories.

Soil survey reports contain definitions of each mapping concept used to delineate a portion of the survey area; these mapping concepts are called map units. The map unit is named for the most extensive soil phases present. In turn, phases are named using both taxonomic concepts (series, usually) and nontaxonomic criteria pertaining to use and management (slope, erosion, etc.). Pedon (profile) descriptions are included. The map unit description (in earlier reports the series description) may contain information on plant rooting depths. Map units that occur together in a predictable pattern commonly are grouped into 5 to 10 so called soil associations; these, in turn, are the map units of the general map for the survey area.

Commonly the number of series and map units pertinent to erosion of cropland of a soil survey is modest. One reason is that many of the map units are either not cropped or erosion is not a problem. If one wishes to reduce the numbers further, one can exclude map units that, in aggregate, occupy a small proportion (about 10%) of the cropland subject to erosion. The numbers remain modest for much larger areas than a single soil survey. A useful concept for multi-survey generalization of soil series and map units is the Major Land Resource Area (MLRA) (Soil Survey Staff, 1965). Within each MLRA, the physiography, climate, and soils are generally similar. Hence, quite similar map unit definitions are employed for the various soil survey areas. Within state portions of MLRA's, map units of different soil surveys have a further affinity because kinds of non-taxonomic criteria (slope, erosion, etc.) and class ranges are usually similar. Consequently, the number of series and map units required for erosion generalizations may not differ greatly between a state subarea of an MLRA and a single survey.

As an example, Table 1 contains the series and map units that encompass 90% of the cropland subject to erosion in the Iowa portion of MLRA 107. For each of the 11 sample soil surveys, we determined the map units suitable for cropland and subject to erosion by water, wind, or both. Map units were included if they were in IIe, IIIe, or IVe subclasses (Klingebiel and Montgomery, 1961), or if the map unit description indicated dominance of cropland subject to erosion from either wind, water, or both. For each sample survey, only the map units necessary to encompass 90% of the cropland area subject to erosion were retained. This was done because for planning purposes on a state MLRA subarea basis complete coverage is unnecessary and by excluding a small portion of the cropland subject to erosion a rather large reduction results in the number of series and map units. The separate lists, each from a survey, were then composited to obtain a single list for the MLRA state subarea. If there were more than a single map unit involving the same soil series, four rules were used to reduce the number of map units in the composite list: (a) retain map units if they differed by two classes in slope, erosion, or both; (b) if three or more classes of slope, erosion, or both were defined, retain the map units with the extremes in slope, erosion, etc.; (c) usually retain map units indicating severe erosion; and (d) retain if properties below the depth of tillage influence or relating to landscape position were in the map unit name. Finally, it should be emphasized that such composite lists pertain only to areas within a state MLRA subarea equal to or larger than the average soil survey for the subarea.

Table 2 expands on Table 1 giving the number of series and map units to encompass 90% of the cropland subject to erosion for 11 state subareas of selected MLRA's. The MLRA's selected have 30% or more cropland in which erosion is the dominant limitation (Stewart et al., 1975, Fig. 7).

DEFINITION OF POTENTIAL ROOTING DEPTH

The first step is to establish the deepest occurrence of either common[3] fine or very fine roots using the class definitions in Soil Survey Staff

---

[3] In strict terms, it is the depth of more than few roots, since root abundance can change from many to few or to none; but for ease of writing, "depth of common roots" is employed.

Table 1. Series and map units that encompass 90% of the cropland subject to erosion in the Iowa portion of Major Land Resource Area 107.†

Series	Surface texture	Slope	Erosion	Other
		%		
Adair	Clay loam	5–9	Moderate	
Bolan	Loam	2–5		
Castana	Silt loam	14–22		
Colo-Judson	Silty clay loam	2–5		
Dempster	Silt loam	2–5		Deep
Dickman	Fine sandy loam	2–5		Benches
Everly	Clay loam	2–5		
Galva	Silty clay loam	2–5		
	Silty clay loam	9–14	Moderate	
Ida	Silt loam	5–9	Severe	
	Silt loam	14–20	Severe	
Judson	Silt loam	2–5		
Judson-Nodaway-Colo‡	--	2–5		
Kennebec	Silt loam	2–6		
Kennebec-Ackmore§	--	2–5		
Ladoga	Silt loam	9–14		
Marshall	Silty clay loam	2–5		
	Silty clay loam	9–14	Severe	
	Silty clay loam	5–9		Mottled subsoil
	Silt loam	2–5		Benches
Mayberry	Clay loam	9–14	Moderate	
McPaul-Kennebec¶	Silt loam	2–5		
Monona	Silt loam	2–5		
	Silt loam	2–5		Benches
	Silt loam	14–20	Severe	
Napier	Silt loam	2–5		
	Silt loam	9–14		
Ocheyedan	Loam	2–5		
Primghar	Silty clay loam	2–5		
Radford	Silty clay loam	2–5		
Sac	Silty clay loam	2–5		
	Silty clay loam	2–5		Loam substratum
	Silty clay loam	2–5		Clay loam substratum
Shelby	Loam	5–9	Moderate	
	Clay loam	14–18		
Storden	Loam	5–15	Moderate	
Terril	Loam	2–5		
Wadena	Loam	2–5		Deep
	Loam	2–5		Moderately deep
Waukegan	Loam	2–5		Moderately deep

† Based on soil surveys of these countries: Cass, Clay, Crawford, Freemont, Harrison, Monona, Page, Plymouth, Sac, Shelby, and Woodbury.
‡ Nodaway part.
§ Ackmore part.
¶ McPaul part.

(1975). Fine roots are 1 to 2 mm in diameter and very fine are < 1 mm. The abundance class for both is 10 to 100 per dm². A value of 25 per dm² is generally accepted by research workers in the field as indicating adequate roots. The value rests on the commonly accepted figure of 0.5 cm of root per 1 cm³ of soil and the generalization by Barley (1970) that the number of intercepts with a plane is half the root length. For soybeans and cotton

Table 2. Number of series and map units that encompass 90% of the cropland subject to erosion for several state subareas of MLRA's.

State	MLRA	No. of soil surveys examined	Number of Series	Number of Map units
Illinois	108	8	45	57
Iowa	104	4	31	31
	107	11	28	40
	108	8	48	62
Kansas	73	4	14	15
	106	5	19	25
Nebraska	106	4	13	23
North Dakota	55	3	42	52
Oklahoma	112	5	12	13
Wisconsin	95	5	63	68
	105	5	31	40
Median	--	--	31	40

the root density should be reduced about 80% from that used for many other crops.[4] If irrigation is needed to grow a commercial crop, use depths of rooting under irrigation. Depths should pertain to plants near physiological maturity. If feasible, the plants should be annuals because their culture renders cropland susceptible to erosion. Preferably the crops should be ubiquitous and common within an MLRA.

Generalizations about plant rooting depths can be obtained from series or map unit descriptions of some soil surveys. For MLRA's where the soil survey has been largely completed and rooting depth information is insufficient, we suggest requesting experienced soil scientists to make rooting depth estimates. Table 3 contains rooting depths assigned in 3 hours by two experienced soil scientists highly conversant with the survey area using prior information alone.

Our objective is to predict the root distribution in reference to the present soil surface after pronounced truncation by erosion. The zone of interest is to 2 m or a root limiting contact if above 2 m. If the base of common roots is unavailable, we begin at the soil surface and continue downward to 2 m in search of a root limiting contact as defined subsequently. We assume that as truncation procedes, common roots may occur at any depth above a root limiting contact. Potential rooting depth, therefore, is considered to be independent of distance from the present soil surface. Moreover, dry conditions at any depth are not considered to be a restriction to rooting.

Three sets of root limiting criteria are given in Table 4. These criteria pertain to most of the major feed and fiber crops of temperate regions. The first set covers taxonomic features as defined in Soil Survey Staff, 1975. Coarse layers are one of these features. It is not clear why roots are impeded at the upper boundary of coarse layers in which particle-size contrasts markedly with the finer soil material above. Coarse soil materials that extend from the soil surface downward do not restrict rooting. We assume that coarse layers are less complete rooting barriers in

---

[4] Personal communication from H. M. Taylor, Iowa State Univ.

Table 3. Rooting depths of corn for representative map units of Dane County, Wisconsin (Glocker and Patzer, 1978).†

Map unit	Base of abundance class‡		Rooting limitations
	Common	Few	
	cm	cm	
Boyer sandy loam,			
2 to 6% slopes	90	125	Sand and gravel
6 to 12% slopes, eroded	80	110	Sand and gravel
12 to 20% slopes, eroded	75	100	Sand and gravel
Griswold loam,			
2 to 6% slopes	95	110	Sand and gravel
6 to 12% slopes	92	105	Sand and gravel
12 to 20% slopes, eroded	75	87	Sand and gravel
Kidder loam,			
2 to 6% slopes	95	120	Sand and gravel
6 to 12% slopes, eroded	80	105	Sand and gravel
12 to 20% slopes, eroded	75	97	Sand and gravel
Plano silt loam,			
0 to 2% slopes	115	150	None
2 to 6% slopes	115	150	None
6 to 12% slopes, eroded	103	146	None
Radford silt loam,			
0 to 3% slopes	90	125	Water table—drained
Rockton silt loam,			
2 to 6% slopes	90	115	Fissured bedrock
6 to 12% slopes, eroded	76	96	Fissured bedrock
12 to 30% slopes, eroded	70	85	Fissured bedrock
Sable silty clay loam,			
0 to 3% slopes	80	105	Water table—drained
Wacousta silty clay loam	50	78	Water table—drained

† Determined by A. J. Klingehoets and G. W. Hudelson.
‡ Classes defined in Soil Survey Staff. 1975.

usually moist or wetter soils. Fehrenbacher (1973) reports appreciable extension of roots into the underlying coarser material in several usually moist soils.

The second set of criteria are nontaxonomic physical criteria indicative of pronounced root restriction. The criteria pertain mainly to strong, dense, slowly pervious lower B and C horizons. The soil material when very moist or wet must lack moderate or strong vertical structural planes at close intervals and in addition exhibit one or more of the following: high pedological strength; high bulk density; or the combination of moderately high strength, low vertical saturated hydraulic conductivity and low linear extensibility. The bulk density limits are generally consistent with studies of root penetration into disaggregated soil material reconstituted to different bulk densities and held at the same water tension (Meredith and Patrick, 1961; Barley et al., 1965; Hemsath and Mazurak, 1974; Cockroft et al., 1969; Taylor et al., 1966; and Gerard et al., 1972).

The third set of criteria involve chemical restriction caused by high neutral-salt extractable Al, low extractable Ca, or both (Brenes and Pearson, 1973; Evans and Kamprath, 1970; Foy, 1974a, 1974b; Howard and Adams, 1965; Kamprath, 1970; Nye et al., 1961). Absolute levels of exchangeable Ca must be very low for expression of Ca deficiency. Very few soils exhibit Ca deficiency independent of Al toxicity. The tentative criterion of 0.02 meq/100 g is based on extractable Ca levels for some of the few soils in the literature to which reference to Ca deficiency is made. A criterion of 0.1 for the ratio of Ca to the sum of bases and Al is based primarily on the work of Howard and Adams (1965).

A ratio of 0.6 between N KCl extractable Aluminum and the sum of extractable bases plus Al ("Al ratio") is used as a criterion for root restriction (Thomas and Cassel, 1979). This Al ratio crudely predicts water soluble Al (Adams and Lund, 1966; Nye et al., 1961; Evans and Kamprath, 1970). Water soluble Al is antagonistic to Ca uptake by plants and also may be injurious directly (Foy, 1974b). Water soluble Al increases sharply above an Al ratio of 0.6, which may explain why this ratio value has proven useful for prediction of root restriction. Activity of the water soluble Al is more critical to root growth than concentration. Activity may be reduced by organic ions (Evans and Kamprath, 1970) and declines as the total ionic concentration rises (Adams and Lund, 1966). These controls over Al activity would act to reduce the value of the Al ratio for root growth prediction.

Soils, species, and cultivars differ in their critical Al ratios (Foy, 1974b; Spain et al., 1975). Brenes and Pearson (1973) reported for two soils found in Puerto Rico that reduction of corn growth began at a ratio of about 0.6, whereas for sorghum growth reduction began at appreciably lower Al ratios. For cotton, Al ratios below 0.3 may indicate toxicity (Adams and Lund, 1966). Adams and Pearson (1967) report minimum toxicity levels of from 0.04 to 0.3 for common field crops.

Adjustments of the depth of the chemical barrier are made using permeability class and extractable Al level. The adjustments assume that ground limestone would be applied to the tillage zone and that Ca dissolved from the limestone would move downward in substantial quantities with the amount for a given moisture regime rising as the permeability increases. In this regard, De Freitas and Van Raij (1975) report downward movement of Ca + Mg at a rate of 0.25 meq/100 g per year for a Latosol of Brazil. Cooke (1967) gives a range of 0.05 to 0.15 for soils of the British Isles. Fox (1980) discusses movement of limestone in soils with variable charge. The rate of translocation can be high but Ca retention by such soils is low. It is recognized that the ameliorative effect of deep leaching of limestone may be much reduced for soils with variable charge. Significance of a given reduction in extractable Al decreases as the initial Al level rises. For this reason, the absolute amount of extractable Al is a consideration.

Appreciable organic matter usually indicates that the horizon is not root-limiting. Therefore, the root-limiting criteria, other than the ones involving extractable Ca and Al, are overridden if the organic carbon in the horizon exceeds 0.6%. Other stipulations are that the soils are in neither fluventic nor pergelic taxa, lack evidence of strong mixing by fauna, and have a postulated genesis that excludes translocation of organic matter as an important process.

Table 4. Criteria for the potential rooting depth.

Kind of limitation	Limitation description†	Reference depth	Illustrative pedons‡
Physical; Taxonomic§	One of the following: 1) Pedogenic horizons, and features such as Duripans, Fragipans, Petrocalcic horizons, Petroferric horizons...; 2) Rock—Lithic and Paralithic contacts; 3) Coarse layers. A change from finer to coarser soil material such as to indicate strong root restriction. The change may be to fragmental soil material of any thickness with no restriction on particle size of the overlying material or sharpness of the transition, or to sandy or sandy skeletal soil materials > 10 cm thick with a laboratory volume water retention difference 33 to 1500 kPa of <0.03, and which together with the overlying soil material meet the requirements for strongly contrasting particle size classes. Two soil moisture regimes are specified: a) The driest extensive soils of the area are usually moist or wetter. b) Extensive soils are drier than usually moist.	Upper boundary Upper boundary Upper boundary if fragmental or moisture regime (b); add 20 cm otherwise.	53, 35, 40 104, 80
Physical; Nontaxonomic	Continuously cemented zone of any thickness, or, a zone > 10 cm thick that when very moist or wet is massive, platy, or for repeat distances < 10 cm, has weak structure of any type and meets at least one of the following criteria where consistence refers to water contents above 33 kPa retention (10 kPa if sandy loam, loamy sand, or sand): 1) Extremely firm or stronger; 2) Have agricultural bulk densities (Mg/m³) equal or exceeding those listed below. Particle-size classes follow family taxonomic classes except that >2 mm is not included.¶ Sandy 1.85 Coarse-loamy 1.80 Fine-loamy 1.70 Coarse-, fine-silty 1.60 Greater than 35% clay (see below)	Upper boundary	11, 96, 113

(continued on next page)

Table 4. Continued.

Kind of limitation	Limitation description†	Reference depth	Illustrative pedons‡
Physical: Nontaxonomic	Clay — Volumetric Water Retention Difference, 33 to 1500 kPa (<2 mm)		
	Pct. 0.05 0.10 0.15 0.20 0.25		
	35–45 1.55 1.55 1.55 1.50 1.40		
	45–55 1.55 1.55 1.45 1.40 1.30		
	55–65 1.55 1.45 1.35 1.30 1.20		
	65–75 1.45 1.35 1.30 1.20 1.15		
	≥75 1.35 1.30 1.30 1.15 1.10		
	3) Very firm and both vertical saturated hydraulic conductivity (permeability) is <10^{-6} m/s (<0.2 inches/hour) and coefficient of linear extensibility (COLE) is <0.03.		
Chemical; Nontaxonomic#	Meets one or more of the criteria to follow: 1) Extractable Ca <0.02 meq/100 g, 2) Extractable Ca divided by the sum of extractable bases (Ca, Mg, Na, K) and Al extractable by $1N$ KCl is <0.1, 3) The ratio of Al extractable with N KCl to the sum of bases plus Al exceeds 0.6.#,††	Upper boundary if criterion 3 applies even if occur below classification control section. Apply criterion 3 applies and Al exceeds 4 me/100 g. Otherwise, add 10 cm to the upper boundary if Ksat <10^{-6} m/s at the upper boundary, 20 cm if 10^{-6} to 5 × 10^{-5} m/s, and 40 cm if >5 × 10^{-5} m/s.	6, 10, 21, 33, 102, 103, 113, 114, 116, 117, 120

† Underlined terms are defined in Soil Survey Staff, 1975. Laboratory measurements are defined in Soil Survey Staff, 1972.
‡ From Soil Survey Staff, 1975.
§ Soil features defined in Soil Survey Staff (1975) that restrict roots markedly. Apply criterion even if occur below classification control section.
¶ Assumes particle density of 2.65 g/cc. Substitute bulk density that gives same void ratio if particle densities differ appreciably from 2.65. Consult footnote to Table 5 for calculation of particle density. Permeability classes similar to those in Klute (1965). Field criteria are described in O'Neal (1952).
# Organic horizons excluded, and also Sulfaquents and Sulfaquepts.
†† No experimental verification known to the writers for spodic horizons or soil material with appreciable volcanic glass. Typical Andepts tend to have low extractable Al.

Table 5. Criteria for the depth change factor.

Property	Criteria	Conventions	Observation	Adjustment
Permeability	Change in dominant phase permeability class between 0 to 30 and 30 to 100 cm.†	Number 1 highest and 7 lowest permeability class. Not applicable if PRD <50 cm.§	1 to 6, 7	None
			1 to 2–5	Increase
			2 to 6	None
			2 to 3–5	Increase
			2 to 1, 7	Decrease
			3 to 2, 4, 5	None
			3 to 1, 5, 6	Decrease
			4 to 3, 5	None
			4 to 1, 2, 6, 7	Decrease
			5 to 2–4, 6	Decrease
			5 to 1, 7	None
			6 to 1, 7	Increase
			6 to 2	None
			6 to 3–5	Increase
			7 to 1	None
			7 to 2–5	Increase
Air-filled porosity	Change among the depth zones D1 = 30–50 cm, D2 = 50–100, D3 = 100–150 in mean air-filled porosity at 33 kPa or a lower tension.	Not applicable if PRD <70 cm or if air-filled porosity in one or more zones is less than 2%. PRD must extend 20 cm into 100–150 cm zone to apply second criterion. Not applicable if texture change within zone necessitates change from 33 kPa to a lower tension for the approximation of field capacity.‡	D1/D2 <0.6 and D2/D3 ≥1.0	Increase
			D1/D2 ≥2.0 or D1/D2 1.1–2.0 and D2/D3 >2.0	Decrease

(continued on next page)

EROSION TOLERANCE FOR CROPLAND 123

Table 5. Continued.

Property	Criteria	Conventions	Observation	Adjustment
Organic carbon	Proportion in the 0 to 30 cm zone of the total from the surface to the PRD or to 0.2% organic carbon, whichever shallower.		Other <25% 25 to 75% ≥75%	None Increase None Decrease
Extractable Ca + K	Location of the depth that encompasses one-half of the Ca + K within the zone 30 cm to the PRD	Apply only if majority of zone from 30 cm to PRD has base saturation by NH_4OAc <50%, and the PRD >100 cm.	Lower 1/3 Middle 1/3 to 2/3 Upper 1/3	Increase None Decrease
Coarse layers	Thickness of sandy or sandy-skeletal material above the PRD, using the particle size definitions in Soil Survey Staff (1975).			Subtract one-half of coarse layer thickness from 200 cm or from the PRD.

† Permeability estimates 0 to 30 cm for the phase usually are not strongly influenced by tillage practice or state in the yearly soil use cycle. Hence, the phase permeability class for the 0–30 cm zone may be employed in criteria for change with depth.

‡ Particle density of 2.65 assumed. Correct the particle density if high in organic carbon assuming 1.4 mg/m³ for organic matter. If high in extractable Fe, correct on the assumptions of a conversion of 1.7 from extractable Fe to the iron minerals present, and a density of 4.1 mg/m³ for the iron minerals. If high in glass of rhyolitic composition, assume sand plus silt has a particle density of 2.3 mg/m³.

§ Current classes in the National Cooperative Soil Survey are as in Klute (1965) with slight differences.

Spodosols and taxonomic intergrades, as well as geographic associates having albic horizons, tend to have high Al ratios in upper B horizons (Soil Survey Staff, 1968-1980). The Al ratio at which root restriction occurs may be higher for these soils than for the Ultisols and associated soils for which the criterion of 0.6 was developed. A possible reason is that generally the horizons of the Spodosols and associated soils with high Al ratios contain more organic carbon which may cause lowered activity to Al in solution (Evans and Kamprath, 1970).

Salts have not been considered as a root limitation in the context here because most soils high in salts have low slopes and are therefore not subject to pronounced water erosion.

T-VALUE ADJUSTMENT FACTOR

The point of departure is that the allowable loss (T value) as currently employed may be separated into two quantities: T value = T_{max} × TAF. The T_{max} would be the T value for soils that have Potential Rooting Depths (PRD) ≥ 200 cm and relatively uniform physical and chemical properties with depth. The T_{max} would be established by people broadly conversant with the soils and agriculture of the area. TAF, the T-value Adjustment Factor, is calculated as follows:

$$TAF = TAF_{50} + (1 - TAF_{50})[(PRD - 50)/150] + DCF$$

Potential Rooting Depth (PRD) is the depth of the root limiting contact or 200 cm whichever is shallower (Table 4). TAF_{50} is the factor for soils with a PRD equal to or shallower than some depth, here 50 cm. The depth below which TAF is constant may be set at depths other than 50 cm. The DCF provides an adjustment for changes with depth above the root limiting contact in properties which indicate sensitivity of productivity to erosion.

Table 5 describes the scheme for obtaining the DCF. It is based on depth-related changes in permeability, air-filled porosity, organic carbon, and the sum of extractable Ca plus K. The occurrence of coarse layers is considered also. Depth limits of 30, 50, 100, and 150 cm are generally used. Thirty centimeters is the assumed depth of tillage influence including traffic compaction. The 50 and 100-cm depths are commonly used to define taxa, and it is advantageous to use them. The 150-cm limit is employed because this is a common sampling depth. Permeability is used because it integrates physical factors affecting plant rooting and values are available for major horizons of all soil series. Change in calculated air-filled porosity at 10 or 33 kPa is employed because many data sets suitable for calculation exist such as those in the Soil Survey Investigation Series; in Holtan et al. (1968), a compendium for soils of experimental watersheds; and in reports of state soil survey laboratories. Air-filled porosity has the advantage of at least indirect measurement whereas most permeability values are estimates. Furthermore, porosity values are continuous and not discrete as is permeability information. Organic carbon is employed because of the wide differences among soils in the proportion of the total organic matter that would be removed by an increment of erosion. Presumably if most of the total organic carbon is within a 30 cm depth, the reduction in productivity would be larger. Extractable Ca plus K is approached analogously to organic carbon, although the uppermost 30 cm is excluded to reduce the effect of amending with ground limestone. Extractable Mg and Na are not considered because their amounts are less directly an indicator of fertility. Coarse layers are introduced because, for most crops, a change from a finer material to sandy or sandy-skeletal material reduces the productivity of the soil.

Depth Change Factor may be zero, positive, or negative. A single negative property change dictates a negative DCF even if other properties are positive. For a positive DCF, at least one measurement must be positive and none of the others negative. The absolute value of DCF is the smaller of either 0.2 or one of the quantities:

positive adjustments: $\frac{1}{2} [1 - \text{TAF (PRD alone)}]$

negative adjustments: $\frac{1}{2} [\text{TAF (PRD alone)} - \text{TAF}_{50}]$

If information is lacking to evaluate a property change, it is assumed that the change is insufficient for either negative or positive identification.

Table 6 contains illustrative TAF values. A TAF_{50} of 0.1 was assumed except when the root limiting contact was produced by high extractable Al or low Ca, in which case TAF_{50} was set at 0.2 because this root limitation is easier to ameloriate than the other kinds.

DISCUSSION

First, a brief review: Our objective is to use the soil survey data base to fashion a scheme for making relative T-value adjustments. The T-value assignment would start with a decision about the maximum allowable loss (T_{max}) to be permitted for soils with no plant rooting restrictions and uniform properties with depth. A T-value Adjustment Factor (TAF) would be calculated by which the maximum allowable loss (T_{max}) is multiplied to give the allowable loss for the particular soil. The factor depends on the depth to a plant rooting limitation with adjustments therefrom depending on changes in soil properties with depth indicative of the effect on soil productivity of erosion. The calculation requires assigning a constant TAF to soils with a plant rooting limitation shallower than a certain depth, here 50 cm.

The calculation of TAF uses criteria that are not plant specific. Species and even cultivars differ in the capability of their root systems to enter and exploit potentially limiting soil horizons. This has been discussed briefly in reference to Al. Another example is the difference among species in response to horizons high in fine-grain carbonate (Unger and Pringle, 1981). We also recognize that other criteria for both root limiting contacts and depth correction factors are feasible. For example, a root limiting contact criterion could be written using percentage of carbonate

Table 6. T-value adjustment factors for selected pedons.†

Illustrative pedon	Potential rooting depth	Adjustment for changes with depth				Coarse layers	T-Value adjustment factor‡
		Permeability§	Air-filled porosity	Organic carbon	Extractable Ca, K		
	cm					cm	
California Hanford 65-027-12 Typic Xerorthent	≥200	None	None	None	NA	None	1.0
California Tehama 64-057-19 Typic Haploxeralf	140	Neg.	None	Neg.	NA	None	0.44
Colorado Kuma 63-058-7 Pachic Argiustoll	112	None	None	None	NA	None	0.47
Hawaii Paaloa 65-007-2 Humoxic Tropohumult	50#	LI	LI	Neg.	NA	None	0.20††
Iowa Marshall 63-015-3 Typic Hapludoll	≥200	None	None	None	NA	None	1.0
Illinois Muren 62-002-2 Aquic Hapludalf	≥200	None	Neg.	Neg.	NA	None	0.80
Indiana Crosby 59-023-2 Aeric Ochraqualf	86	Neg.	Neg.	None	NA	None	0.21
Kentucky Crider 59-005-2 Typic Paleudult	<200	Neg.	None	Neg.	NA	None	0.80
Louisiana Calloway 63-042-2 Glossaquic Fragiudalf	64	LI	LI	Neg.	NA	None	0.14

(continued on next page)

Table 6. Continued.

Illustrative pedon	Potential rooting depth	Adjustment for changes with depth					T-Value adjustment factor‡
		Permeability§	Air-filled porosity	Organic carbon	Extractable Ca, K	Coarse layers	

Michigan
  Marlette 70-029-2
  Glossaboric Hapludalf
	86	None	None	Neg.	NA	None	0.26

Minnesota
  Racine 70-020-1
  Typic Eutrochrept
	≥200	None	None	Neg.	NA	None	0.80

Mississippi
  Savannah 61-048-3
  Typic Fragiudult
	58	LI	LI	Neg.	NA	None	0.12

Nebraska
  Mayberry¶ 58-067-5
  Aquic Argiudoll
	≥200	None	None	None	NA	None	1.0

New Hampshire
  Scituate 59-007-3
  Typic Fragiochrept
	86	None	None	Neg.	NA	None	0.21

New Jersey
  Downer 58-008-7
  Typic Hapludult
	≥200	None	None	Neg.	None	39	0.57

New York
  Langford 63-006-3
  Glossic Fragiudalf
	46	NA	NA	NA	NA	None	0.10

Oklahoma
  Newtonia 60-058-2
  Typic Paleudoll
	86	None	None	None	NA	None	0.32

Puerto Rico
  Catalina 59-010-8
  Tropeptic Haplorthox
	106	None	None	None	Neg.	None	0.41††

(continued on next page)

Table 6. Continued.

Illustrative pedon	Potential rooting depth	Adjustment for changes with depth					T-Value adjustment factor[‡]
		Permeability[§]	Air-filled porosity	Organic carbon	Extractable Ca, K	Coarse layers	
Tennessee Armour 60-094-23 Ultic Hapludalf	≥200	None	None	Neg.	NA	None	0.80
Texas Burleson 62-043-3 Udic Pellustert	≥200	None	None	Pos.	NA	None	1.0
Texas Crockett 62-129-3 Udertic Paleustalf	145	Neg.	None	None	NA	None	0.47
Wisconsin Hochheim 68-008-1 Typic Argiudoll	104	None	LI	None	NA	None	0.42
Wyoming Apron 65-007-12 Typic Torriorthent	≥200	None	None	Neg.	NA	29	0.63

[†] All pedons in the soil survey investigations series except the one from Tennessee, which is in Edwards et al., 1974. NA means not applicable given the conventions in Table 5, and LI means lack of information.
[‡] Carried to one more place than justifiable for a soil phase to facilitate calculation checks.
[§] Applying phase class placements to pedons for illustrative purposes.
[¶] Published as Adair.
[††] TAF_{50} = 0.2.
[#] Assumes more than few roots to 30 cm.

<0.05 mm, structural expression, and bulk density. Similarly, a criterion could be written, analogous to the one for the coarse layers in Table 5, to account for strong restriction of roots between large structural units at water contents below field capacity. A criterion for clayey horizons might be a bulk density of 1.6 g/cc exclusive of the crack space produced from drying to 200 kPa. In our judgment, however, it would be premature to propose more criteria now.

We have presented a detailed set of instructions because only such a scheme can be tested. We think that the problem here is similar to that posed in the development of the USDA soil taxonomy system and is subject to solution in the same fashion by the repeated construction, application, and reconstruction of detailed schemes. We recognize that a major determinant for T-value adjustment should be potential tilth. An early future step should be to combine an approach such as herein with a scheme for tilth prediction.

Finally, we believe that decisions about T-value adjustments should be made by a small group of people intimately conversant with an agricultural area, perhaps a Land Resource Region (Soil Survey Staff, 1965). These individuals should be charged to use criteria such as presented but with wide latitude for exercise of judgment.

LITERATURE CITED

Adams, F., and Z. F. Lund. 1966. Effect of chemical activity of soil solution aluminum on cotton root penetration of acid subsoils. Soil Sci. 101:193-198.

————, and R. W. Pearson. 1967. Crop response to lime in the southern United States and Puerto Rico. *In* Soil Acidity and Liming. R. W. Pearson and F. Adams (ed.) Agron. Mono. 12 Am. Soc. Agron., Madison, Wis. 174 p.

Barley, K. P. 1970. The configuration of the root system in relation to nutrient uptake. Adv. Agron. 22:159-201.

————, D. A. Farrell, and E. L. Greacen. 1965. Influence of soil strength on the penetration of a loam by plant roots. Aust. J. Soil Res. 3:69-79.

Brenes, E., and R. W. Pearson. 1973. Root responses of three gramineae species to soil acidity in an Oxisol and an Ultisol. Soil Sci. 116:295-302.

Cockroft, B., K. P. Barley, and E. L. Greacen. 1969. Penetration of clays by fine probes and root tips. Aust. J. Soil Res. 7:333-348.

Cooke, G. W. 1967. The control of soil fertility. Crosby Lockwood, London. 526 p.

De Freitas, L. M. M., and B. Van Raij. 1975. Residual effects of liming a Latosol in Sao Paulo, Brazil. *In* E. Bornemisza and A. Alvarado (ed.) Soil management in tropical America. Soil Sci. Dep. North Carolina State, Raleigh. 565 p.

Edwards, M. J., J. A. Elder, and M. E. Springer. 1974. The soils of the Nashville Basin. Tenn. Agric. Exp. Stn. Bul. 499. 123 p.

Evans, C. E., and E. J. Kamprath. 1970. Lime response as related to percent Al saturation, solution Al, and organic matter content. Soil Sci. Soc. Am. Proc. 34:893-896.

Fehrenbacher, J. B., B. W. Ray, J. D. Alexander, and H. J. Kleiss. 1973. Root penetration in soils with contrasting textural layers. Illinois Res. 15:5-7.

Fox, R. L. 1980. Soils with variable charge: Agronomic and fertility aspects. *In* Soils with variable charge. B. K. G. Theng (ed.). New Zealand Soc. Soil Sci., Lower Hutt, New Zealand. 448 p.

Foy, C. D. 1974a. Effects of soil calcium availability on plant growth. *In* E. W. Carson (ed.) The plant root and its environment. Univ. Press Virginia, Charlottesville. 691 p.

————. 1974b. Effects of aluminum on plant growth. *In* E. W. Carson (ed.) The plant root and its environment. Univ. Press Virginia, Charlottsville. 691 p.

Gerard, C. J., H. C. Mehta, and E. Hinojosa. 1972. Root growth in a clay soil. Soil Sci. 114: 37–49.

Glocker, C. L., and R. A. Patzer. 1978. Soil survey of Dane County, Wisconsin. U.S. Dep. Agric., Washington, DC. 293 p. w/90 inset sheets.

Hemsath, D. L., and A. P. Mazurak. 1974. Seedling growth of sorghum in clay-sand mixtures of various compactions and water contacts. Soil Sci. Soc. Am. Proc. 38:387–390.

Holtan, H. N., C. B. England, G. P. Lawless, and G. A. Schumaker. 1968. Moisture-tension data for selected soils on experimental watersheds. ARS 41-144. 609 p.

Howard, D. D., and F. Adams. 1965. Calcium requirement for penetration of subsoils by primary cotton roots. Soil Sci. Soc. Am. Proc. 29:558–562.

Kamprath, E. J. 1970. Exchangeable aluminum as a criterion for liming leached mineral soils. Soil Sci. Soc. Am. Proc. 34:252–254.

Klingebiel, A. A., and P. H. Montgomery. 1961. Land—Capability classification. SCS, USDA, Agric. Handb. no. 210. 21 p.

Klute, A. 1975. Laboratory measurement of hydraulic conductivity of saturated soil. *In* C. A. Black (ed.) Physical methods of soil analysis. Am. Soc. Agron., Madison. 770 p.

Meredith, H. L., and W. H. Patrick, Jr. 1961. Effects of soil compaction on subsoil root penetration and physical properties of three soils in Louisiana. Agron. J. 53:163–167.

Nye, P., P. Craig, N. T. Coleman, and J. L. Ragland. 1961. Ion exchange equilibria involving aluminum. Soil Sci. Soc. Am. Proc. 25:14–17.

O'Neal, A. M. 1952. A key for evaluating soil permeability by means of certain field clues. Soil Sci. Soc. Am. Proc. 16:312–315.

Soil Survey Staff. 1965. Land resource regions and major land resource areas of the United States. USDA Handb. No. 296. 82 p.

————. 1972. Soil survey laboratory methods and procedure for collecting soil samples. SCS, USDA, Wash. Soil Survey Invest. Rep. No. 1.

————. 1968–1980. Soil Survey Laboratory data and descriptions for some soils of New England States (Rep. 20), New York (Report 25), Minnesota (Report 33), Wisconsin (Report 34), Michigan (Report 36).

————. 1975. Soil taxonomy, a basic system of soil classification for making and interpreting soil surveys. Agric. Handb. no. 436. USDA, SCS.

————. 1979. Map of status of soil surveys. 1 Jan. 1979. USDA, SCS.

Spain, J. M., C. A. Francis, R. H. Howeler, and F. Calvo. 1975. Differential species and varietal tolerance to soil acidity in tropical crops and pastures. *In* E. Bornemisza and A. Alvorado (ed.) Soil management in tropical America. Soil Sci. Dep., North Carolina State, Raleigh. 565 p.

Stewart, B. A., D. A. Woolhiser, W. H. Wischmeier, J. H. Caro, and M. H. Frere. 1975. control of water pollution from cropland. Vol. I. Report No. EPA-600/2-75-026a or ARS-H-5-1. 111 p.

Taylor, H. M., G. M. Roberson, and J. J. Parker, Jr. 1966. Soil strength-root penetration relationships for medium- to coarse-textured soil materials. Soil Sci. 102:18–22.

Thomas, D. J., and D. K. Cassel. 1979. Land-forming Atlantic coastal plains soils: Crop yield relationships to soil physical and chemical properties. J. Soil Water Conserv. 34: 20–24.

Unger, P. W., and F. B. Pringle. 1981. Pullman soils—distribution, importance, variability, and management. Texas Agric. Ext. Serv. Bul. B-1372.

Chapter 11

Improved Criteria for Developing Soil Loss Tolerance Levels for Cropland[1]

TERRY J. LOGAN[2]

ABSTRACT

The present national objective of erosion control in the USA is to maintain the productivity of the soil resource and still allow for its maximum utilization. For agricultural land use, this objective is applied at the state and regional level as the soil loss tolerance (T) value; it is based on soil interpretation at the series level and ranges from 2 to 11 metric tons/ha/year (1 to 5 tons/acre/year). Used with the Universal Soil Loss Equation (USLE), it serves as a guide to farmers for selecting conservation practices suited to their soil and crop management conditions. The primary criterion currently used for establishing T values is topsoil thickness such that deeper soils can tolerate higher levels of annual erosion than shallower soils. This criterion is examined and several additional factors are considered: soil structure, existing erosion conditions, specific rooting requirements of crops, available erosion control technology, and non-point source pollution control objectives.

THE T VALUE CONCEPT

In the decades following the initiation of erosion control research in the USA, great strides have been made in the estimation of long-term erosion from agricultural land. The Universal Soil Loss Equation (USLE) developed by Wischmeier and his associates (USDA Agricultural Handbook 537, 1978) has been widely used as a means of assessing long-term soil loss and to predict the effectiveness of various conservation and crop management practices. The USLE has more recently been used to assess soil

---
[1] Contribution of the Ohio Agric. Res. and Development Center, Wooster 44691.
[2] Associate Professor, Agronomy Dep.

Copyright © 1982 ASA, SSSA, 677 South Segoe Road, Madison, WI 53711. *Determinants of Soil Loss Tolerance.*

loss from urban and forested areas, and even to predict erosion from individual storm events. Wischmeier (1975) has cautioned against the careless misapplication of the USLE beyond its intended use, but nevertheless, it remains the most useful tool currently available for assessment of soil erosion.

The USLE has been used to determine the conservation and crop management practices required to keep annual soil loss within some allowable or tolerable level, the soil loss or tolerance value (T value) which is the subject of this symposium. In its simplified form, the USLE is given as:

$$A = R \times K \times LS \times C \times P \qquad [1]$$

R = rainfall erosion index
LS = factor which combines the effects of slope length and steepness
K = soil erodibility factor
C = cover and crop management factor
P = supporting conservation practices factor
A = Annual soil loss (metric tons/ha)

A more detailed description of the equation and the derivation of its factors are given in USDA Agricultural Handbook 537 (1978).

The factor combination RKLS is a measure of the intrinsic susceptibility of a soil on a particular landscape to erode, and to reduce the annual soil loss, A, to the tolerance value, it is necessary to choose some combination of crop management (C) and conservation (P) practices according to the relationship:

$$CP = T/RKLS \qquad [2]$$

The farmer is free to choose the CP combination which is best suited or most economically feasible for his area. In cases where the required CP combination is high enough, it can be achieved by the use of cover management only, i.e., reduced tillage or crop rotation. In many areas, however, the ratio of A/T is high enough that additional conservation practices are required.

The objective of applying the soil loss tolerance concept is an obvious one, i.e., preserving our soil resource, but a systematic or scientific rationale for establishing individual T values has not been developed. Bender (1962), in the proceedings of an SCS workshop gave three general criteria for establishing T values:

1. Soil loss should be reduced to a level which will maintain an adequate soil depth favorable for crop and timber production over a long period of time. An evaluation should be made of the effect of soil erosion on crop yields.

2. Soil losses should be held below those that cause severe gullying in the field or cause serious siltation in waterways, terrace channels, drainage ditches, road ditches, etc.

3. Plant nutrient losses should not be excessive.

These are not really criteria at all, however, but merely fairly broad ob-

Table 1. Soil Conservation Services guidelines for T values (SCS, 1977).

Depth criteria (cm)	T value (metric tons/ha/year)				
	11.2 (5)†	9.0 (4)	6.7 (3)	4.5 (2)	2.2 (1)
> 100 to Bedrock	X				
> 100 to Sand and/or gravel	X				
50–100 to Bedrock		X			
50–100 to Sand and/or gravel		X			
50–100 to Fragipan		X			
50–100 to Claypan		X			
25–50 to Sand and/or gravel			X		
25–50 to Bedrock				X	
10–50 to Claypan			X		
< 50 to Fragipan			X		
< 25 to Bedrock					X
< 25 to Sand and/or gravel				X	
< 10 to Claypan					X

† Tons/acre.

jectives. The only consistent criterion used by SCS in establishing T values is favorable rooting depth[3]. This is illustrated in Table 1.

SOIL RENEWAL ASSUMPTIONS

The T-value concept is based on the premise that soil is being renewed by weathering of parent material and by aeolian and fluvial deposition, and that we can reasonably estimate these renewal rates. The concept also recognizes that the rate of soil renewal is usually less than the erosion rate and therefore, accepts that there will be some continued loss of the nation's soil resource. We are in essence mining the soil in order to produce food and fiber in the same way that we mine our coal resource. The Soil Conservation service suggests, through the T value, rates at which this resource may be depleted without seriously impairing the utility of the remaining soil base for 200 years. Therefore, as given in Table 1, those soils with the deepest topsoil horizons are permitted to erode more rapidly than shallower soils. By so doing, however, we are accepting that many of our best soils will be progressively deteriorated faster than our less productive soils. Since most estimates of soil renewal are 0.5 metric tons/ha/year (< 0.2 tons/acre) (Logan, 1977), and T values range from 2 to 11 metric tons/ha/year (1 to 5 tons/acre) (SCS, 1977), one can assume that, in time, our agricultural soils will have progressively less favorable rooting depth. Even more significantly, rooting depth thickness of our deepest soils will begin to approach those of our shallower soils. At that point, given the present criteria, more restrictive T values would be imposed. The very important question of actual rates of topsoil renewal will be discussed by others in this symposium, but two points are worth

[3] USDA-SCS. 1977. Midwest Technical Service Center. TSC Advisory Soils-LI-13, 14 July.

mentioning here: (1) rates of rock weathering are not necessarily similar to rates of soil renewal and (2) whatever the rates of soil renewal are, they are probably lower than present T values. Soil renewal at the parent material/soil interface, however, may be considerably slower than rock weathering as the system is much closer to equilibrium than weathering rock. On the other hand, if soil renewal implies an increase in organic matter and nutrients, there is evidence that this can be achieved much more rapidly than rock weathering. Dickson and Crocker (1953, 1954) found that carbon and N accumulated on an exposed California mudflow much more rapidly than changes in physical characteristics, and these rates were considerably faster than rock weathering.

Topsoil Removal and Crop Production

During the post-Dust Bowl era of accelerated research on erosion, a number of regional studies were conducted to determine the impact of topsoil loss on crop production (Stallings, 1957). Many of these studies involved incremental removal of topsoil and treatment of the exposed subsoil with lime, fertilizer, or manure. One such study at Wooster, Ohio showed that subsoil corn yields were 17% of those where no topsoil was removed, and were 46 and 79% after 4 and 11 years, respectively (Ohio Exp. Stn. Handb., 1957). The acid subsoil (Fragiudalf) responded to additions of lime and phosphate, and after 13 years total N content had increased from 1,053 to 1,893 kg/ha. This however, was still only about 50% of the original N content of the topsoil. A similar study on a finer-textured soil failed to give similar responses to chemical and manure additions, and the poor physical condition of the exposed subsoil apparently contributed to the lack of yield response. This points out that replacement of nutrients and pH control, while simple enough to attain, do not guarantee that crop production will be increased since poor physical conditions may prove to be the limiting factor to optimum crop production.

Loss of Organic Matter and Nutrients by Erosion

During runoff, nutrients are lost from the soil as soluble constituents (primarily NO_3-N) or attached to the sediment (P, exchangeable cations, trace elements and organic N and S). Loss of organic matter affects both the nutrient levels of the soil and its physical condition.

A soil loss of 1 metric ton/ha (0.5 tons/acre) will result in a reduction of topsoil depth of about 0.008 cm, and a loss of 13, 3.5, and 1 kg/ha of K, P, and N, respectively, for a medium-textured soil, assuming that the sediment that leaves the field has the same nutrient content as the original topsoil. However, during erosion and sediment transport, there is selective removal of the finer particles and humus, together with soluble nutrient forms. The sediment, therefore, is enriched with nutrients and organic matter. As erosion increases, the degree of enrichment decreases

Table 2. Mean annual soil loss and N, P, and K loss per ton of soil loss from several rotations in the period 1962–1971 (Burwell et al., 1975).

Rotation	Soil loss	N	P	K
	metric tons/ha		kg/metric tons of soil loss†	
Fallow	37.00	4.06	0.90	0.23
Continuous corn	16.47	4.73	1.13	0.23
Corn-rotation	7.54	4.77	1.15	0.25
Oats-rotation	4.35	5.42	1.21	0.45
Hay-rotation	0.02	205.00	34.00	228.00

† Total nutrient loss (soluble plus particulate) divided by soil loss.

as coarser particles are eroded and transported, and the sediment begins to have the same nutrient and organic matter content as the original surface soil (Porter, 1975). Therefore, erosion control does not reduce nutrient loss proportionally. In addition, as soil loss decreases, a greater proportion of the removed nutrient is in the soluble form. This is illustrated by the work of Burwell et al. (1975) for various cropping systems (Table 2). The data is expressed here as nutrient loss per unit of soil loss, and shows that, as soil loss is reduced to < 5 metric tons/ha, nutrient loss is not reduced proportionately. This is due to the increased enrichment of P and N in the sediment as well as a greater proportion of total nutrient loss in soluble forms as erosion decreases. This was also shown by Johnson et al. (1979) in Iowa where the decrease in erosion with a till-plant system compared to the conventional plow tillage was much greater than the corresponding decrease in total N and P, and the proportion of N and P lost in the soluble form increased with decreasing erosion. These findings imply that reducing nutrient loss may not be a significant justification for reducing soil loss, especially when one considers that nutrient inputs are still a relatively small part of modern agricultural production costs.

Deterioration of desirable soil structure as topsoil is eroded is obvious to anyone who tries to plow over the eroded nobs in farm fields. The Soil Conservation Service recognizes this in their criteria (Table 1) for T values. Lower T values are given for soils with subsoil characteristics that are undesirable for rooting, including physical characteristics such as high bulk density and low water holding capacity. Deterioration of soil structure is difficult to quantify and it is equally difficult to determine the effects of poor structure on crop production. However, undesirable subsoil structure is likely to be more limiting than nutrients since the later can be built up with additions; soil structure needs to be considered more strongly in establishing T values.

Achievable Levels of Soil Loss Reduction

The previous discussion would indicate that there are several aspects of the present concept of T value which require careful examination. One is our acceptance that our soil resource will continue to erode faster than soil renewal rates even if we meet present T values. Secondly, our willing-

Table 3. Universal Soil Loss Equation estimates of gross erosion in the Lake Erie Drainage Basin with several soil management strategies (LEWMS, 1979).

	Present conditions	Meet T	Spring plow	Fall plow	Mulch tillage	No-till
			metric tons/ha/year			
Maumee River Basin, Ohio	7.8	4.3	7.4	8.3	3.4	1.1
Cattaraugus Creek, New York	7.8	6.0	7.8	9.0	3.8	1.3
Lake Erie basin (U.S. side only)	7.2	4.3	6.7	7.6	3.1	1.1

ness to accept erosion of our better soils at faster rates than poorer soils. Thirdly, the lack of hard scientific data to support the range of 1 to 5 tons/acre (2 to 11 metric tons/ha) of tolerable soil loss presently used in the USA.

In establishing a national policy of soil loss abatement, we should also consider the ability of farmers to reduce soil loss with currently available soil management technology. No-till and conservation tillage systems can reduce annual soil loss to 2 metric tons/ha (<1 ton/acre) in many instances (Burwell et al., 1975), and when used on moderately well and well-drained soils, can even result in increased crop yields (Bone et al., 1977). Lake Erie Wastewater Management Study (LEWMS, 1979) used USLE analysis to show that application of conservation tillage or no-till in computer simulations could reduce soil losses in the Lake Erie drainage basin below that which could be obtained by meeting the T value for all soils in the Basin (Table 3). They assumed that mulch tillage (chisel plow) and no-till would be applied only on those soils which gave no crop yield reduction with reduced tillage management. These are usually better-drained, sloping soils and represent some of the more erosive soils in the area. The farmer here is clearly capable of reducing soil loss to less than T with little or no economic hardship and may even realize higher yields and lower energy costs.

Soil Loss Tolerance and Water Quality

Throughout this paper, I have deliberately avoided discussion of the off-farm impacts of erosion and impaired water quality and the costs associated with these impacts since they will be addressed by others in the symposium. However, it is tempting to use the T value, a ready-made, accepted strategy for soil loss reduction, to meet the further goal of improved water quality. While the broad objectives of erosion control and nonpoint source pollution control are similar, there are several important differences: (1) while erosion control is targeted to a particular soil (series) wherever it may occur, water quality objectives are usually developed at the watershed level, and the soils in the basin requiring treatment are those which have the greatest impact on water quality. A soil with moderate erosion hazard adjacent to a stream may contribute more to the pollutant load to that stream than a highly erosive soil situated some distance from the nearest waterway. These critical area, or hydrologically active, soils may require soil loss reduction below the T value (Johnson and Berg, 1979). In this case, the two objectives should be kept separate and financial assistance or other additional support be used to reduce soil loss from the T value for agricultural production maintenance down to that required to meet environmental quality objectives.

T-Value Criteria at the Local Level

Soil loss tolerance values are determined by SCS at the state and national levels by soil series, and are based primarily on favorable rooting depths. This classification, however, does not consider many local conditions. Rooting depth requirements of local crops may vary considerably. A surface soil thickness of 30 to 45 cm over an acid clay subsoil might be adequate for corn-soybean production but marginal for alfalfa. Chemical as well as physical subsoil characteristics should be considered in light of specific crop needs.

The T-value criteria at the soil series level also fail to consider local variation in existing erosion conditions. The T-value of 5 tons/acre (11.2 metric tons/year) for a soil with >100 cm of favorable rooting depth (Table 1) would no longer apply to the specific situation where erosion has reduced the thickness below the 100 cm criterion. This could occur in areas where soil survey was completed decades ago and erosion has been severe since then.

In light of these considerations, it would appear that the concept of applying a single value of soil loss tolerance by soil type is unjustifiable. Considerations such as downstream water quality, crop rooting depth requirements and local effectiveness of erosion control practices all argue for a local or regional approach to setting T.

I would propose the following approach in establishing T values for our nation's soils:

1. Maintain the present system which would represent an upper limit of 2 to 11 metric tons/ha (1 to 5 tons/acre) at the soil series level and could be modified as our knowledge of the effects of long-term erosion improve.

2. Reduce the upper limit of 11 metric tons/ha (1 to 5 tons/acre) for those soils and geographic areas where lower levels can be attained with slight shifts in soil management.

3. Soil Conservation Service state and district personnel would modify T values based on local crop production needs and experiences with soil conservation technology.

4. Soil loss limits would be established on a distributed basis to meet local or regional water quality objectives and the additional costs of meeting soil loss limit rather than T could qualify for assistance programs.

5. T-values would be reassessed periodically to reflect actual erosion conditions on soils previously surveyed.

LITERATURE CITED

Bender, W. H. 1962. Soil erodibility and soil loss tolerance. *In* Soil loss prediction for the North Central States. Chicago Workshop, SCS.

Bone, S. W., D. M. Van Doren, Jr., and G. B. Triplett, Jr. 1977. Tillage research in Ohio. A guide to the selection of profitable tillage systems. Ohio Coop. Ext. Serv. Bull. 620.

Burwell, R. E., D. R. Timmons, and R. F. Holt. 1975. Nutrient transport in surface runoff, as influenced by soil cover and seasonal periods. Soil Sci. Soc. Am. Proc. 39:523-528.

Dickson, B. A., and R. L. Crocker. 1953. A chronosequence of soils and vegetation near Mt. Shasta, California. II. The development of the forest floors and the carbon and nitrogen profiles of the soils. J. Soil Sci. 4:142-155.

————, and ————. 1954. A chronosequence of soils and vegetation near Mt. Shasta, California. III. Some properties of the mineral soils. J. Soil Sci. 5:173-192.

Handbook of Ohio experiments in agronomy. 1957. Ohio Agric. Exp. Stn., Wooster.

Johnson, H. P., J. L. Baker, W. D. Shrader, and J. M. Laflen. 1979. Tillage system effects on sediment and nutrients in runoff from small watersheds. Trans. ASAE. 22:1110-1114.

Johnson, M., and N. Berg. 1979. A framework for nonpoint pollution control in the Great Lakes Basin. J. Soil Water Conserv. 34:68-74.

Lake Erie Wastewater Management Study. 1979. Phase II Report. Army Corps of Engineers, Buffalo District, Buffalo, N.Y.

Logan, T. J. 1977. Establishing soil loss and sediment yield limits for agricultural land. *In* Soil erosion and sedimentation. Proc. Natl. Symp. Soil Erosion and Sedimentation by Water. ASAE, Chicago, Ill.

Porter, K. S. 1975. Nitrogen and phosphorus, food production, waste and the environment. Ann Arbor Science, Ann Arbor, Mich. 372 p.

Stallings, J. H. 1957. Soil Conservation. Prentice Hall, Inc. p. 195-220.

Wischmeier, W. H. 1975. Use and misuse of the universal soil loss equation. J. Soil Water Conserv. 31:5-10.

————, and D. D. Smith. 1978. Predicting rainfall erosion losses—a guide to conservation planning. USDA/SEA. Agric. Handb. 537. 58 p.

Chapter 12

Economics of Soil Erosion Control with Application to T Values[1]

JOHN F. TIMMONS AND ORLEY M. AMOS, JR.[2]

ABSTRACT

Pressure on existing cropland, emanating from a complex set of factors increasing demand for agricultural products and limiting agriculture's production response, is accelerating erosion losses. These losses take place on existing cropland and particularly on cropland converted from noncrop uses. Public concerns regarding effects of increasing soil erosion losses on future productivity of soils and environmental quality of water are being answered with legislation and agency actions to limit soil erosion. These actions require establishment and enforcement of erosion limits of tolerance termed T values.

Establishment and enforcement of T values affect farm income, U.S. capacity to export crops, availability and prices of food for domestic and foreign consumers as well as future soil productivity and quality of water and air. Consequently, all these factors and effects must be considered in establishing particular T values. T values will likely vary due to soil differences and the nature and magnitude of external effects.

---

[1] Presented at annual meeting of Soil Sci. Soc. of Am., Fort Collins, Colorado. 8 Aug. 1979. Journal Paper No. J-9625 of the Iowa Agric. and Home Economics Exp. Stn., Ames, Iowa. Project No. 2045.

[2] Professor of economics and Charles F. Curtiss distinguished professor in agriculture, Iowa State University, and assistant professor of economics, Oklahoma State Univ., formerly Research Associate, Iowa State University, respectively.

The authors express their sincere appreciation to W. D. Shrader and Pierre Crosson for their advice and assistance in preparing this paper, to John G. DeMabior for calculating all metric conversions and to Resources for the Future for helping fund this research. Of course, the authors alone must accept full responsibility for what is written as well as what is omitted from the paper.

Copyright © 1982 ASA, SSSA, 677 South Segoe Road, Madison, WI 53711. *Determinants of Soil Loss Tolerance.*

T value effects on net farm income were estimated by extending and adjusting a programming model currently used in estimating conversion of noncropland to cropland in Iowa. Income and erosion effects were estimated from 14 various T values under several combinations of crop prices, factor costs, cropping practices, and discount rates. Effects of various T values on present value of net income ranged from $179 to $1,706 per ha ($72 to $690 per acre) under one set of variables and from $2,076 to $6,752 per ha ($840 to $2,732 per acre) under another set of variables in the data set concerning conversion of noncropland to cropland.

Further analysis in estimating relevant effects of T values preceding and as an essential part of their establishment, invites close working relationships between soil scientists, agronomists, and economists in conducting the needed research.

INTRODUCTION

Increasing demands with higher prices for agricultural products since 1972 coupled with constraints upon agriculture's ability to meet these demands are accelerating soil erosion and associated deterioration of water and air qualities. Growing public concerns over soil productivity and water and air quality are generating legislative and administrative actions to abate soil erosion losses and sedimentation of watercourses.

These public concerns and their resultant state and federal actions are bringing limits, including maximum limits on nonpoint pollution from soil erosion. These actions are being taken by state departments of environmental quality and the federal Environmental Protection Agency. States are establishing soil and water conservancy districts and other means of controlling soil erosion. Congress recently held hearings (26-27 July 1979) on the soil erosion control problems and possible legislative actions.

All of these concerns and actions necessitate establishment of limits on soil erosion. Enforcement of such limits exerts profound effects, not only on future soil productivity and environmental quality, but also upon the financial integrity of farmers, capacity of the nation to export agricultural products, and the availability of farm products for domestic consumers as well as the product prices which consumers must pay. Consequently, care and caution should be exercised in the establishment of erosion tolerances whose effects will be experienced throughout local, state, and national economies and extend well into the international scene.

State and federal agencies are increasingly inclined to establish universal erosion loss limits without varying the limits according to the needs of particular soils and erosion effects. Once limits are set, they tend to become fixed and unchangeable. Consequently, establishment of such limits should be taken seriously and based upon rigorous analysis of relevant and accurate information.

The purpose of this paper is to outline possible economic contributions (1) to the understanding of soil erosion problems and (2) in analyzing levels of erosion control as an inherent part of the process of evaluating and establishing T values representing maximum soil losses to be tolerated.

SOIL AND PEOPLE

Interactions between soil and people are ancient and remain important. People could not exist without soils, but soils could exist without people. Soil erosion existed long before interactions with people and developed spectacular configurations of landscape as well as some of our most productive upland soils through wind erosion and most of our productive valleys and deltas through water erosion.

Considerable discussion has occurred over the years regarding effects of soil and other natural resources on people as compared to effects of people's use of these resources on soils and other resources. According to Ely and Wehrwein (1940, p. 26),

"Some writers carry the influences of physical factors beyond the mere influence of climate, etc. on productive powers. They claim that man's life is determined by his physical environment, and can even see differences in the institutions, actions, and opinions of men as the result of living on different soil types. Others swing to the opposite direction; while not denying the relation of man to his physical universe, they place the emphasis on man as the determining factor of his own social and economic life."

Ellsworth Huntington (1924) in his *Civilization and Climate* published over a half century ago, emphasized the influence of climate, soils, and other natural resources upon human progress. More than a century ago, George P. Marsh (1864), in his prophetic book *Man and Nature*, emphasized effects of man on soils and other natural resources. He concluded that man along with water and wind becomes an important factor changing the nature of his natural environment including soils, frequently with greater destructive force than all the natural forces put together. Kellogg (1937, p. 142) reasoned that

"On the natural landscape all factors are interdependent. There is a relationship between soil and vegetation, between soil and climate, between climate and vegetation, between soil and parent rock, between soil and slope, and even between climate and slope, but all of these factors cooperate in the production of an actual soil. The soil is the final synthetic expression of all of these factors working together, and by which the nature of landscape can be characterized better, more directly and completely, than any other factor or combination of factors."

Kellogg (1935) also emphasized the effects of people on soil erosion in the dust storms of the 1930s. He reasoned that people could redress these effects on soil erosion through strip farming and other land use practices for which people are responsible. In other words, people could damage the soil, but they also could prevent or repair the damages within limits.

In their use of natural resources, people necessarily disturb them both in terms of what is extracted from them as well as what is put into them. Also, the technologies used in conducting these necessary disturbances are crucial to the productivity and life of soils and other natural resources.

Unlike our fossil-fuel resource base for energy, which is exhaustible and nonrenewable, our soil and water resource base for food, feed grains, and fiber is potentially exhaustible, but it can be made renewable for the most part, if and only if soil and water resources are used and managed carefully.

It appears that recent food and feed demands generated by exports with higher prices, and thus providing greater incentives for farmers to produce, are exacting serious soil and water losses through erosion and sedimentation of watercourses. In effect, we are exporting our soil and water quality in the form of food and feed grain exports.

In the soil and people equation, important interactions exist. Soils affect people and people affect soils. In affecting soils, people can use investments and technology that deplete and erode soils, or they can use investments and technology that maintain or enhance productivity and prevent erosion. The question arises as to level of productivity to be maintained and the level of soil loss to be tolerated.

CURRENT AND PROSPECTIVE FACTORS ASSOCIATED WITH SOIL EROSION INCREASES

For more than four decades, federal, state, and local efforts including education, technical assistance, cost sharing arrangements, and monetary grants, have been devoted to the reduction of soil erosion losses. While these measures have seldom attained stated goals, increases in soil erosion were lessened, and occasionally soil erosion losses were decreased.

Despite all these efforts and experiences with erosion control measures, erosion losses during the 1970s remained excessive and, in numerous instances, have actually increased. This turnaround in erosion losses has been associated with a complex combination of important and intransigent factors that may well continue indefinitely into the future.

Two sets of factors are involved. One set is associated with increases in demand for agricultural products. The other set is associated with limitations on agriculture's capacity to respond to demand increases.

Among the factors associated with increased demand for agricultural products is increased per capita income in developed and developing countries. Increased per capita income with its effect upon kind and amount of food demanded has derived largely from increased national income and/or improved income distribution within industrial nations and the OPEC countries plus the countries that they assist with grants and loans. Of course, increased population as propounded by Malthus (1798) and his long line of successors continues to put pressure on demand for agricultural products but per capita income presently seems to be more important. Food and feed grain exports to the Communist countries exert important influences on increased demand for agricultural products. Another demand-increasing factor is the need for foreign exchange by the United States to purchase petroleum, vehicles, electronics, textiles, and other products which until recently were produced domestically. Agricultural exports are being relied upon increasingly by the United States as a matter of public policy to help balance our international payments. They have quadrupled in dollar value during the past decade and will most likely exceed $48 billion during the 1981 export year ending 30 June 1981. The continuing imbalance of payments, however, puts pressure on the United States to increase further and substantially the export of agricultural products in the national interest of the integrity of the dollar and inflation control. At the same time, the continuing excess of imports over exports reduces the value of the dollar in terms of other currencies benefitting from our trade imbalance and therefore adds to the demand of our relatively low dollar priced U.S. agricultural products in terms of these other currencies.

Another set of concomitant factors of natural and manmade origins adversely affects the ability of U.S. agriculture to respond to increased demands for its products. One of these factors, weather conditions, particularly inadequate moisture during the middle 1970s, interfered with crop production throughout large sections of the nation. Second, petroleum based inputs, particularly fertilizers, pesticides and energy, responsible for much of the increase in crop yields in recent decades, are becoming increasingly expensive and scarce. Third, environmental quality constraints, especially nonpoint water pollution including suspended silt and chemical residuals, discourage continuation of some of the more productive agricultural practices. Fourth, interferences with agricultural production emanating from expanding urban and industrial development into productive farming areas frequently conflict with and aggravate crop production practices because of noise, odors, chemicals, and tractor machinery traffic on rural roads affecting recent urban settlers. Fifth, there appears to be a leveling off of productivity increases per unit of land due to the above factors and possibly due also to inadequate production research of basic and applied nature.

Sixth, reduction in yield increases (per unit of land) resulting from all these factors necessitates bringing additional land into cultivation. These additional land units usually are less productive than current cropland, hence output per unit of land is less, particularly in terms of net value productivity. Also, additional land converted to cropland frequently is more fragile and susceptible to water and wind erosion with adverse effects on environmental quality and in the continuing productivity of soil.

Soil erosion increases arising from these two sets of factors are being exacerbated by other deeply imbedded factors, which have remained unmitigated during the past four or five decades.

Of course, no one can project the future nature and extent of these trends with any substantial degree of certainty. However, Cory and Timmons (1978) estimated erosion losses through 1985 from a base period of 1969–1971. Under an assumption of increased export demand, there was a 28% increase in planted cropland and a 72% increase in soil erosion losses for the 12 Corn Belt states. These losses vary from a 106% increase in Iowa to a 40% increase in Illinois.

Future behavior of these factors would appear to bring continuing pressures on our soils with further erosion-increasing consequences.

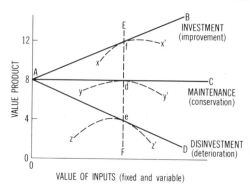

Fig. 1. Illustration of value products per value of fixed total input at three levels of productivity for a particular soil.

CONCERNS AND CONTRIBUTIONS OF ECONOMICS

Major economic concerns about soil erosion levels center upon (1) disinvestment in soils affecting their productivity and (2) external diseconomies created by erosion with effects both offsite and over time.

DISINVESTMENT IN SOILS

Soil erosion affects soil productivity as illustrated in Fig. 1 developed after Timmons et al. (1961). Soil productivity, measured in terms of value product, is depicted by the three curves, AB, AC, and AD. AB represents investment, AD represents disinvestment, and AC represents maintenance of soil productivity. These three concepts are similar to improvement, conservation, and depletion or deterioration of soils, respectively, in the vocabulary of soil scientists.

Curves xx', yy¾, and zz' represent production functions along which inputs may be varied back and forth, usually in accordance with prices and costs, yielding various amounts of value product. The vertical dashed line, EF, runs through the optimum production points f, d, and e of the production functions under investment (AB), maintenance (AC), and disinvestment (AD). With this example, starting with point A, curve AD represents disinvestment from soil erosion by the extent of d to e, which represents a drop from eight to four, or four units of value product. Similarly, starting at point A, curve AB represents an increase of productivity from d to f, which yields an increase in value product from 8 to 12, or four units of value product.

With this illustration, conservation could be defined as shifting from point d to point f or as the prevention of shifting from point d to point e. Assumptions made under this hypothetical illustration are: soils considered as homogeneous, technology held constant, and prices of product, and costs of inputs held constant. Of course, these assumptions may be relaxed by incorporating changes into the analysis.

From this conceptual framework, economic effects of various amounts of soil erosion losses may be estimated by analytical procedures.

EXTERNAL DISECONOMIES CREATED BY SOIL EROSION

In addition to on-site effects of soil erosion treated above, soil erosion can create off-site and overtime effects, which are costly but which costs are not reflected in the use, practices, and output of soils on which erosion occurs. These effects are known as external diseconomies. These two effects are termed interspatial and intertemporal, respectively. These concepts are developed and discussed extensively in economics literature by Bator (1958), Brewer (1971), Coase (1960), Mishan (1971), and others. If these external costs are not borne by the land user, his motivation tends toward overuse of soils, with consequences in the form of soil erosion costs that are shifted to other people and other entities residing in other areas and/or in subsequent time periods.

Consequently permissible soil erosion loss limits must include these external effects as well as on-site productivity effects.

Included within these external effects are delivery ratios of suspended silt in watercourses resulting from erosion losses. In addition to delivery ratios, the suspended silt serves as a transport agent for residuals through adsorption and absorption actions. All these factors are relevant in analyzing and in establishing tolerance values for soil erosion control.

AN ANALYTICAL FRAMEWORK FOR TESTING T VALUES: MODEL AND APPLICATION

Specific T values may be tested for effects on both on-site value productivity and external diseconomies through programming. However, neither sufficient data nor time were available to make these tests for inclusion in this paper. Consequently, limited tests on varying T values were made by using the data base and model developed by Amos (1979)[3] in a recent study estimating potential supply of cropland in Iowa. This study placed a 11.2 metric ton/ha (5 tons/acre) per year constraint on the conversion of noncropland to cropland under varying product prices, factor costs, discount rates, and technologies. Erosion losses were estimated through application of the Universal Soil Loss Equation (USLE) developed by Wischmeier and Smith (1965) and Wischmeier (1976).

From the data base and procedures in the Amos study (1979), two changes were introduced in the computer runs for this paper. First, instead of a soil loss constraint of 11.2 metric tons/ha (5 tons/acre) per year, soil losses were varied in an effort to ascertain how net income streams might be affected by varying T values. Differences between present value of net income streams for the several T values programmed as constraints constitute estimates of costs incurred by farmers in meeting soil loss constraints.

---

[3] Amos, O. M., Jr. Supply of Potential Cropland in Iowa. Unpublished Ph.D. Dissertation. Graduate College, Iowa State University, 1979.

Although these above results should indicate economic implications for farmers operating under various T value constraints, the results provide little or no indication of societal impacts of external diseconomies.

Requirements for decreasing erosion losses usually bring farmers increasing costs, and the benefits are shared with society.[4] It is extremely difficult to measure the social costs of soil erosion. However, in an attempt to provide some indication of the effects that alternative T values might have on society beyond income effects on farm income, a second change was made in the data and in the model.

This approach was pursued by comparing estimated gross soil losses under each of the T values specified and by examining the amount of land converted to cropland in each of five categories. These categories are intended to reflect disinvestment and external diseconomies of soils included in relatively homogeneous categories.[5] These five categories, including 246 soils, are as follows:

    I. Erosion and disinvestments are not problems. Soils in this category are alluvial, colluvial, or flat.

   II. Soils in which the principal soil property changed as a result of erosion is loss of organic matter and on which productivity of an eroded soil can be restored to about the same level as an uneroded soil by addition of plant nutrients, principally N and P. Soils in this category are medium textured throughout and are of a depth greater than the rooting depth of the crop in question.

  III. Soils on which plant nutrients supplied in fertilizer form do not completely restore the productivity of eroded soils. Eroded soils in this category that are adequately fertilized will yield 60 to 90% as much as similar comparably treated uneroded soils. Further yield increases can be achieved only at considerable expense which is not economic under present cost-price conditions.

  IV. Soils on which yields are severely (50 to 100%) reduced by erosion even when adequate plant nutrients are supplied. Like category III soils, further yield increases can be achieved only at an uneconomic cost.

   V. Land made unfit for cultivation by erosion. Soils in this category are usually unfit for pasture, also.

---

[4] Of course, increased costs to farmers resulting from lower levels of tolerated erosion ultimately will be passed on to consumers through higher prices of farm products. This occurs because of the relatively competitive conditions of agricultural production, elasticity of response to changing costs, and elasticity of demand for farm products. Thus, agricultural production may be expected to decline in the short run, with increasing costs generated by compliance with decreasing erosion loss limits, resulting in relatively higher prices to consumers. In the long run, consumers receive benefits in the form of increased supplies and relatively lower costs of farm products, assuming that decreased soil losses result in improved productivity over what would have occurred without decreased soil losses. Hence, one might reason that consumers pay costs and bear the effects resulting from either erosion control or the lack of it. This reasoning may be used as an argument for public erosion control payments as required since citizens would pay the costs either as taxpayers contributing to erosion control costs or as consumers paying higher food prices without erosion control.

[5] Assistance of Dr. William Shrader, Dep. of Agronomy, Iowa State University, is greatly appreciated in developing these categories.

Examples of soils included in each category are shown in Appendix B.

By examining the portion of each soil category contained in the estimate of potential cropland in the Amos (1979) study, an indication of external diseconomies and disinvestments can be estimated.

For a given soil-loss constraint, discount rate, and set of crop prices, the quantity of potential cropland was estimated with a benefit-cost framework. First, all noncropland in Iowa, so identified by the USDA, Soil Conservation Service (SCS) National Erosion Inventory (1977), was divided into 300 relatively homogeneous classes.[6] This grouping was based on soil type, slope phase, and land use so as to identify variations in productivity, soil erosion, and costs of converting noncropland to cropland between each class of land.

For each of the 300 land classes, the costs and benefits of converting it from noncropland to cropland were calculated. The costs included: (1) investment expenditures needed to make the land suitable for cropping such as drainage tile installation, tree removal, or terracing and (2) opportunity costs of revenue foregone from the current noncropland use (e.g., pasture or forest land) after it is converted to cropland.

Five crop rotations were introduced and tested: (1) continuous corn, (2) corn-soybean, (3) corn-soybean-corn-oats-meadow-meadow, (4) corn-oats-meadow-meadow-meadow-meadow, and (5) continuous meadow. Tillage and erosion control practices introduced and tested were (1) conventional moldboard plowing, (2) minimum tillage, (3) straight row, (4) contouring, and (5) terracing. Benefit-cost ratios were calculated for each combination of practices; e.g., corn-soybean, straight row and moldboard; corn-soybean, straight row and residue; continuous corn, contouring, residue, etc. In all, 30 benefit-cost ratios were calculated for each land class. The most profitable, i.e., the largest benefit-cost ratio, combination of practices was selected for use in further analysis.

Another criterion used to determine potential cropland within each class was average soil loss. For each benefit-cost ratio calculated, the average soil loss resulting from the combination of practices was estimated by application of the USLE (Wischmeier and Smith, 1965; Wischmeier, 1976). For each land class to be deemed potential cropland, it had to satisfy both the benefit-cost criterion and also a soil-loss constraint. That is, average soil loss must be less than a specified maximum. If all 30 benefit-cost ratios indicated that a land class was potential cropland, but average soil loss exceeded the specified limit for all 30 sets of practices, the land class was not considered potential cropland.

Using the soil-loss constraint criterion enables some of the economic impacts of alternative soil-loss constraint to be identified. Two variables calculated in this respect were gross soil-loss and net income. Gross soil-loss is the total soil loss from potential cropland meeting both criteria. Net income is the total benefits minus all costs for potential cropland.

---

[6] The 300 noncropland classes are a further division of the 246 soils mentioned earlier. The two classifications differ in two aspects. First, the 300 classes pertain only to noncropland. Several of the 246 soils did not appear in the National Erosion Inventory in noncropland uses. Second, the 300 noncropland classes were dimensioned by slope phase, or level of previous erosion, the 246 soils were not. The 300 noncropland classes are a finer, but not exhaustive, subdivision of the 246 soils.

Under one set of conditions composed of a particular soil-loss constraint, discount rate, and set of prices, one estimate of a specific combination gross soil-loss and present value of net income was obtained. Alternative soil-loss constraints, discount rate, and crop prices produce different results. Comparison of results can give insight into effects of each combination of conditions.

Two sets of crop prices and three discount rates, 4, 6, and 8% were used. In simplifying procedures, 1977 costs were used in all the scenarios. Four crops and their respective prices were used in the analysis: corn, soybeans, oats, and hay. The initial set of crop prices, set A, was average crop prices for 1977, with corn at $0.08 per ka ($2.08 per bushel), soybeans at $0.26 per kg ($7.05 per bushel), oats at $0.09 per kg ($1.34 per bushel), and hay at $62.48 per metric ton ($56.68 per ton). The second set of crop prices, set B, was approximately double that of set A. For set B the prices were corn at $0.16 per kg ($4.05 per bushel), soybeans at $0.50 per kg ($13.74 per bushel), oats at $0.18 per kg ($2.61 per bushel), and hay at $121.75 per metric ton ($110.45 per ton). Prices for 1977, set A, were used to test the soil-loss constraint against current economic conditions. However, less than 121,406 ha (300,000 acres) of potential cropland were estimated with set A. Since this was only 5% of the 2.43 million ha (6 million acres) of noncropland used in the analysis, set B crop prices also were employed, which identified 2.02 million ha (5 million acres) of potential cropland.

In making changes in the data, model, and runs for testing various T values, 13 alternative soil-loss constraints were introduced in the analysis. Maximum permissible average soil losses of 0, 2.2, 4.5, 6.7, 9.0, 11.2, 13.4, 15.7, 17.9, 20.2, 22.4, 33.6, and 44.8 metric tons/ha per year (0, 1, 2, 3, 4, 5, 6, 7, 8, 9, 10, 15, and 20 tons/acre per year) were tested, in addition to an unconstrained solution.

Appendix A Table 1 records estimates of present value of net income per acre (per ha) under both sets of crop prices and three discount rates for the 13 soil-loss constraints and the unconstrained solution. For price set A, present value of net income ranges from $179 to $1,706 per ha ($72 to $690 per acre). For set B the range is from $2,076 to $6,752 per ha ($840 to $2,732 per acre). This difference between the sets arises because set B crop prices are double set A, thus increasing the benefits without changing costs. Cost remained at 1977 levels.

Present value of net income per ha is graphed against the 13 soil-loss constraints for each discount rate and both sets of prices in Fig. 2 and 3, respectively. As the soil loss constraint is relaxed, average net income increases. However, there are segments of the curves in Fig. 2 and 3 that deserve explanation. Between 0 and 1 ton per acre per year (0 and 2.2 metric tons/ha per year) of soil erosion, present value of net income has a distinct jump. This indicates that reducing soil loss to zero has a large impact on farmers' incomes. A second characteristic noticeable in all six curves is the near vertical portion between 9.0 and 20.2 metric tons/ha per year (4 and 9 tons/acre per year). Evidently, once erosion control investments are made, they are effective over a considerable range of soil loss control.

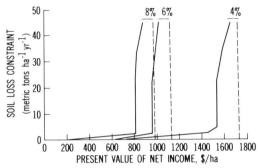

Fig. 2. Present value of net income from potential cropland for alternative soil loss constraints under price set A and three discount rates.

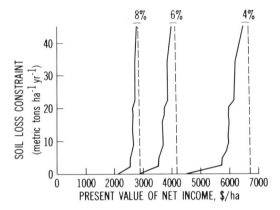

Fig. 3. Present value of net income from potential cropland for alternative soil loss constraints under price set B and three discount rates.

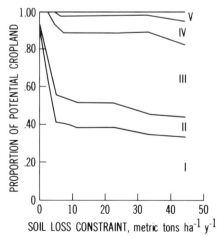

Fig. 4. Proportion of potential cropland by soil category for alternative soil loss constraints, price set B, 4% discount rate.

The dashed line drawn next to each curve represents the average net income from the unconstrained solution wherein no costs are incurred to control soil erosion. As might be expected, relaxing the soil-loss constraint permits average net income to approach the unconstrained average net income.

A second set of results, which may help provide insight into the benefits of reduced soil erosion, is presented in Appendix A Table 2. All noncropland identified by the SCS and incorporated in the study was divided into the five categories discussed earlier and based on potential disinvestment from soil erosion.[7] Appendix A Table 2 presents the proportion of potential cropland falling in each soil category, for the 4% discount rate.

Figure 4 shows the relative proportions for set B crop prices. This graph indicates that, as the soil-loss constraint is relaxed, relatively less good land is used and relatively more poor land is used. That is, up to 44.8 metric tons/ha per year (20 tons/acre per year) or soil-loss constraint, soil categories III, IV, and V, which incur some disinvestment, continually increase their relative share of potential cropland. This indicates that disinvestment is more likely, with the more relaxed soil-loss constraint.

COMPLEMENTARIES BETWEEN SOIL SCIENCE AND ECONOMICS

Throughout this paper, the need for cooperation between researchers in soil science and economics has become obvious in the analysis of soil erosion effects and the establishment of T values to constrain soil erosion. Research by the economist is no more reliable or relevent than the data obtained from soil scientists. Research by the soil scientist remains unfinished for decision-making and policy purposes until and unless analyzed and interpreted in terms of its effects and implications regarding farmers, consumers, the national economy and foreign trade.

Progress in research, with its reservoir of accumulated theories and data and its complex methodologies, has far exceeded the universal doctor stage of an Aristotle. The individual scholar possesses neither the comprehension nor the years of life to assimilate the vast array of information accumulated since Aristotle. This matter was anticipated by Marsh in his *Man and Nature* (1864) over a century ago. It was fully recognized by Kellogg in his "Soil and the People" (1937) more than four decades ago.

Consequently, scientists, as specialists and scholars within their respective divisions of knowledge termed disciplines, must cooperate in research wherein the nature of the problem and its solution transcends the limits of individual disciplines. And most of the important problems require the talents and services of two or more disciplines. Such is the case with soil erosion, which involves soils and people in their use of soils.

---

[7] See Appendix B for examples of soils included in each category.

Appendix A. Tables Summarizing Programming Results

Table 1. Present value of average net income from potential cropland for alternative soil loss contraints.

	Price set A†			Price set B‡		
Discount rate =	4%	6%	8%	4%	6%	8%
Soil loss constraint (metric tons/ha)			$ per hectare			
0	722	635	179	4,672	2,937	2,076
2.2	1,420	948	807	5,748	3,571	2,556
4.5	1,513	948	807	5,748	3,571	2,556
6.7	1,513	948	807	5,873	3,642	2,598
9.0	1,513	948	807	5,971	3,724	2,660
11.2	1,513	948	807	5,973	3,726	2,660
13.4	1,513	948	807	5,977	3,728	2,664
15.7	1,513	948	807	5,977	3,728	2,664
17.9	1,513	948	807	6,001	3,760	2,671
20.2	1,513	948	807	6,027	3,773	2,676
22.4	1,513	948	807	6,188	3,883	2,749
33.6	1,564	993	821	6,277	3,922	2,767
44.8	1,645	1,029	875	6,589	4,114	2,864
Unconstrained	1,706	1,118	975	6,752	4,221	2,931

† Assumes average crop prices for 1977. Source [Amos, 1979].
‡ Assumes crop prices about two times 1977 level. Source [Amos, 1979, p. 108].

Table 2. Proportion of potential cropland within each soil category for alternative soil loss constraints under 4% discount rate.

	Price Set A†					Price Set B‡				
Soil category =	I	II	III	IV	V	I	II	III	IV	V
Soil loss constraint (metric tons/ha)					%					
0	97.2	1.1	1.7	--	--	93.2	1.5	5.3	--	--
2.2	84.0	4.2	10.9	0.9	--	62.1	12.6	25.3	--	--
4.5	83.9	4.4	10.8	0.9	--	41.6	13.5	37.9	7.0	--
6.7	83.9	4.4	10.8	0.9	--	40.9	13.0	35.8	8.1	3.3
9.0	83.9	4.4	10.8	0.9	--	39.8	13.3	36.1	8.0	2.8
11.2	83.9	4.4	10.8	0.9	--	38.4	13.0	37.8	8.0	2.8
13.4	83.9	4.4	10.8	0.9	--	38.3	13.1	37.9	8.4	2.3
15.7	83.9	4.4	10.8	0.9	--	38.3	13.1	37.9	8.4	2.3
17.9	83.9	4.4	10.8	0.9	--	38.3	13.1	37.9	8.4	2.3
20.2	83.9	4.4	10.8	0.9	--	38.3	13.1	37.9	8.4	2.3
22.4	83.9	4.4	10.8	0.9	--	38.3	13.1	37.9	8.4	2.3
33.6	83.9	4.4	10.8	0.9	--	35.1	11.2	42.3	9.3	2.1
44.8	83.9	4.4	10.8	0.9	--	33.8	9.7	39.1	13.1	4.3
Unconstrained	82.6	4.2	10.3	1.8	1.1	32.6	9.5	38.5	14.2	5.2

† Assumes average crop prices for 1977. Source [Amos, 1979].
‡ Assumes crop prices about two times 1977 level. Source [Amos, 1979, p. 108].

Appendix B. Some examples of major soils contained in each of five soil categories.

	Slope (%)
Category I. All alluvial and colluvial soils, soils of flats and depressions and miscellaneous land types not subject to erosion.	
Canisteo silty clay loam	--
Colo silty clay loam	--
Haynie silt loam	--
Nodaway silt loam	--
Webster silty clay loam	--
Zook silty clay loam	--
Category II. Soils subject to erosion because of favorable subsoils.	
Galva silt loam	5–9
Ida silt loam	9–14
Marshall silty clay loam	5–9
Monona silt loam	5–9
Category III. Erosion damage results in more than loss of fertility but subsoils are usually profitable to farm.	
Clarion loam	2–5
Grundy silty clay loam	2–5
Kenyon loam	2–5
Shelby loam	5–9
Tama silty clay loam	5–9
Category IV. Erosion damage results in serious loss of productivity even with adequate fertility.	
Adair clay loam	5–9
Clarinda silty clay loam	5–9
Seymour silt loam	2–5
Weller silt loam	5–9
Category V. Land made unfit for cultivation because of erosion	
Dubuque silt loam	9–14
Gosport silt loam	9–14
Sogn loam	9–14

LITERATURE CITED

Bator, F. M. 1958. The anatomy of market failure. Q. J. Econ. 72:351–379.

Brewer, M. F. 1971. Agrisystems and ecoculture: Or: Can economics internalize agriculture's environmental externalities? Am. J. Agric. Econ. 53:848–858.

Coase, R. H. 1960. The problem of social cost. J. Law Econ. Vol. III:1–44.

Cory, D. C., and J. F. Timmons. 1978. Responsiveness of soil erosion losses in the corn belt to increased demands for agricultural products. J. Soil Water Conserv. 33:221–226.

Ely, R. T., and G. S. Wehrwein. 1940. Land economics. Macmillan Co., New York.

Huntington, Ellsworth. 1924. Civilization and climate. Yale University Press. 3rd ed. New Haven.

Kellogg, Charles. 1937. Soil and the people. Annals of the Association of American Geographers.

Kellogg, C. E. 1935. Soil blowing and dust storms. USDA Misc. Pub. 221, USDA, G.P.O., Washington, DC.

Malthus, T. R. 1798. An essay on the principle of population, 1798. Reprinted by Penguin Books, Baltimore, 1970.

Marsh, G. P. 1864. Man and nature, physical geography as modified by human action. Schribners Publishing Co., New York.

Mishan, E. J. 1971. The postwar literature on externalities: An interpretive essay. J. Econ. Lit. 9:1-28.

Timmons, J. F., Marion Clawson, T. W. Edminster, Luna Leopold, R. J. Muchkenhirn, H. A. Steele, and N. J. Volk. 1961. p. 9-11. *In* Principles of resource conservation policy with some applications to soil and water resources. National Academy of Sciences, National Res. Coun., Pub. 885, Washington, DC.

U.S. Department of Agriculture, Soil Conservation Service. National Erosion Inventory. 1977. Magnetic Tape, Statistical Laboratory, Iowa State University, Ames, Iowa.

Wischmeier, W. H., and D. D. Smith. 1965. Predicting rainfall—erosion losses from cropland east of the Rocky Mountains. Agric. Handb. no. 282.

―――. 1976. Use and misuse of the Universal Soil Loss Equation. J. Soil Water Conserv. p. 5-9.